Richard A. Nixon
x 486

**World's
Air Forces**

Also by David Wragg
World's Air Fleets

World's Air Forces

David W. Wragg

Published in the United States
by HIPPOCRENE BOOKS, INC

Published in England by
Osprey Publishing Ltd, Reading, Berkshire
© Copyright 1971 Osprey Publishing Ltd
All rights reserved

SBN 85045 038 1

Printed by The Berkshire Printing Co Ltd, Reading, England

Contents

List of plates
Preface
Abu Dhabi
Afghanistan
Albania
Algeria
Argentina
Australia
Austria
Belgium
Bolivia
Brazil
Bulgaria
Burma
Cambodia
Canada
Central African Republic
Ceylon
Chad Republic
Chile
Republic of China (Taiwan)
Chinese People's Republic
Colombia
Democratic Republic of the Congo (Kinshasa)
Congo (Brazzaville)
Cuba
Czechoslovakia
Dahomey
Denmark
Dominican Republic
Ecuador
Egypt (United Arab Republic)
Eire
El Salvador
Ethiopia
Finland
France
Gabon
Federal German Republic
German Democratic Republic
Ghana
Greece
Guatemala
Guinea Republic
Haiti
Honduras
Hong Kong
Hungary
India
Indonesia
Iran
Iraq
Israel
Italy
Ivory Coast
Jamaica
Japan
Jordan
Kenya
Republic of Korea
Democratic People's Republic of Korea
Kuwait

Laos
Lebanon
Libya
Malagasy Republic
Malaysia
Mali Republic
Mauritania
Mexico
Mongolia
Morocco
Muscat and Oman
Nepal
Netherlands
New Zealand
Nicaragua
Niger
Nigeria
Norway
Pakistan
Paraguay
Peru
Philippines
Poland
Portugal
Rhodesia
Rumania
Saudi Arabia
Singapore
Somali
South Africa
South Yemen Republic
Spain

Sudan
Sweden
Switzerland
Syria
Tanzania
Thailand
Togo
Tunisia
Turkey
Uganda
Union of Soviet Socialist
  Republics
United Kingdom of Great
  Britain and Northern
  Ireland
United States of America
Uruguay
Venezuela
Vietnam (South)
People's Republic of Vietnam
  (North)
Yemen
Yugoslavia
Zambia
Captions to the Plates
Modern Aircraft Types
Major Defence Agreements
  and Alliances
NATO identification names
  given to Russian aircraft
Abbreviations used in the
  text

List of Plates

1. BAC Canberra B.62
2. De Havilland Canada DHC-6 Twin Otter
3. Dassault Mirage III-O
4. Westland S-58 Wessex
5. Grumman S-2E Tracker and Douglas A-4G Skyhawk
6. SAAB 105E
7. Sud Alouette III
8. Sikorsky S-65S
9. Dassault Mirage 5
10. Dornier Do 27 and Sud Alouette II
11. Grumman HU-16 Albatross
12. De Havilland Canada DHC-5 Buffalo
13. Douglas C-47 Dakota
14. North American F-86F Sabre
15. Dassault Mirage F.1 and Dassault Mirage G
16. Dassault Mirage IV
17. Rheims Aviation Max Holste 1521M Broussard
18. Sud SA.330 Puma
19. McDonnell Douglas F-4 Phantom II
20. Dassault-Breguet-Dornier Alpha Jet
21. Dornier Do 27
22. Lockheed F-104G Starfighter
23. Hawker Siddeley Gnat
24. BAC Canberra B(1)58
25. Fiat G.91
26. Agusta-Bell 204B Iroquois
27. NAMC XC-1A
28. Lockheed F-104J Starfighter
29. BAC 167 Strikemaster
30. Sikorsky S-61A
31. Breguet Br.1150 Atlantic
32. Fokker F-27M Troopship
33. Sukhoi Su-7B
34. Mikoyan MiG-21PF 'Fishbed D'
35. Mikoyan MiG-17
36. Aermacchi MB.326K
37. Hawker Siddeley Shackleton MR.3
38. Lockheed C-130B Hercules
39. Dassault Mirage III-BZ
40. Dassault Mirage III-RS
41. Tupolev Tu-20 'Bear' and General Dynamics F-102 Delta Dagger
42. Kamov Ka-25 'Hormone'
43. BAC Lightning F.53
44. Panavia 200 Panther
45. BAC-Breguet Jaguar

46 Hawker Siddeley Harrier
47 Hawker Siddeley Hunter
48 Hawker Siddeley Nimrod
49 Westland SH-3D Sea King
50 Westland Wasp
51 Westland-Bell 47 Sioux
52 Westland Scout
53 Westland WG.13 Lynx
54 McDonnell Douglas F-15
55 General Dynamics F-111A
56 Lockheed C-5A Galaxy
57 Northrop F-5A Freedom Fighter and Northrop T-38 Talon
58 Cessna A-37
59 Douglas A-1E Skyraider, Sikorsky CH-3B and Lockheed KC-130H Hercules
60 Cessna O-2A, Bell UH-1D Iroquois and Sikorsky S-58
61 Sikorsky CH-3C
62 Ling-Temco-Vought A-7E Corsair II
63 McDonnell Douglas F-4B Phantom II
64 Grumman E-2 Hawkeye
65 Lockheed P-3 Orion
66 Kaman UH-2 Seasprite
67 Bell AH-1J Sea Cobra
68 Hawker Siddeley Harrier AV-8
69 McDonnell Douglas F-4 Phantom II
70 Boeing-Vertol CH-47 Chinook

Preface

Of all the many books on aircraft, I think there are none which give comprehensive information about the owners and operators of military aircraft. WORLD'S AIR FORCES is intended to fill this gap by giving brief historical details of every military air force in the world including army, navy and marine air arms, with as much information as possible regarding their existing and planned strength. I have also provided a brief note on the major defence alliances, and a list of the most important aircraft types in service. Historical details include important dates and events in the development of the air forces and air arms, with summaries of the aircraft operated at each stage of their history, and reference to the more important actions in which they took part. Information relating to existing strengths includes numbers of aircraft and numbers of squadrons, together with a figure for the total number of personnel.

This amounts to 152 entries covering the military aviation of 112 nations. There are also seventy photographs, and details of 208 major aircraft types, although these are necessarily brief since this is a book about the user, not the equipment.

I wish to thank the Central Office of Information, the Fleet Air Arm Museum, Heeresfilm und Lichtbildstelle, and all the aviation, military and naval attachés, the government departments and aircraft manufacturers who have so generously supplied photographs and information for this book. Without their help it could not have been written.

DAVID W. WRAGG

Air Forces

ABU DHABI

Air Wing, Abu Dhabi Defence Forces

In common with the other Trucial States, including Muscat and Oman, Abu Dhabi has in recent years built up her own defence forces, with British assistance, in anticipation of Britain's withdrawal from the Persian Gulf. The British withdrawal was planned for 1971, but will now be postponed, while much also depends on whether the Trucial States can form themselves into a federation. Air Wing equipment includes ten Hawker Siddeley Hunter fighter-bombers with two conversion trainers, four de Havilland Canada DHC-4 Caribou and two Britten-Norman BN-2 Islander transports, and four Agusta-Bell 206 JetRanger helicopters.

AFGHANISTAN

Royal Afghan Air Force

Afghanistan's history of military aviation dates from 1924, when an army air arm was formed with two Bristol F.2B fighter-reconnaissance aircraft which were flown by German pilots. These were soon supplemented by a squadron of R-2 reconnaissance-bombers, a gift from the Soviet Union which also trained Afghan pilots. The entire force of aircraft was destroyed during a civil war in 1928–9.

The Royal Afghan Air Force was formed in 1937 with orders placed for eight Hawker Hart bombers, sixteen Meridionali Ro 37 reconnaissance aircraft and eight Breda Ba.25 trainers, which were all delivered the following year. British and Italian instructors established a flying school, while the RAF in India also trained some Afghan personnel. In 1939 the RAfAF received twenty Hawker Hinds, some of which remained in service until 1957, while five out of twelve Avro Anson 18s first introduced to Afghan service in 1948 remained in service for almost twenty years!

In 1955, Afghanistan concluded an aid agreement with the Soviet Union under which the RAfAF was to receive Mikoyan MiG-17 fighters, Ilyushin Il-28 jet bombers and Il-14 transports, Antonov An-2 utility transports and SM-1 (Polish-built Mil Mi-1) helicopters. These were delivered two years later, with MiG-15UTI and Yakovlev Yak-18 trainers. Russian instructors and personnel were seconded to Afghanistan, while some Afghan personnel were also trained in India and the USSR. Airfields were built by both the United States and the Soviet Union.

Currently, Mikoyan MiG-21 interceptors have replaced the seven squadrons of MiG-17 fighters; three bomber squadrons operate forty-five Ilyushin Il-28s; while some twenty-five Ilyushin Il-14 and five Il-18 transports are also operated. Six Mil Mi-1 and eighteen Mi-4 helicopters, plus Yak and MiG trainers, complete the RAfAF's aircraft strength. The RAfAF is still an integral part of the Afghan Army.

ALBANIA

Albanian People's Army Air Force

Albania is the smallest East European nation, and one of the poorest. An attempt was made in 1914 to form an Albanian Air Corps, but this proved a false start since the aircraft were taken over by the Austrian Army when World War I commenced in 1914. Financial difficulties foiled inter-war attempts to form an air force, and in April 1939 the country was occupied by Italian forces.

Soviet troops occupied Albania at the end of World War II. The Albanian People's Army Air Force was formed in 1947 with a gift of twelve Yakovlev Yak-3 fighters and a few Polikarpov Po-2 biplane trainers from the Soviet Union, with Soviet personnel seconded as instructors and advisers. In 1955 Albania was a founder member of the Warsaw Pact, and received the Army Air Force's first jet equipment, Mikoyan MiG-15 fighters and MiG-15UTI trainers. In recent years, however, Albania has taken a very independent line regarding the Soviet Union, and is most usually allied with Communist China, from where much of her military equipment now comes.

Current equipment of the 2,500-strong APAAF includes a squadron of ten Shenyang F-6 (Chinese-built MiG-19) fighters, three squadrons of Shenyang F-4 (MiG-17) fighters, and two squadrons with a total of twenty MiG-15s. There is also a transport squadron with three Ilyushin Il-14 and three Antonov An-2 aircraft, and a helicopter squadron with a couple of Mil Mi-1 and eight Mi-4 helicopters. Training equipment consists of Yakovlev Yak-18 and Mikoyan MiG-15UTI aircraft.

ALGERIA

Algerian Air Force *Force Aérienne Algérienne*

Algeria became independent of France in 1962, and immediately started to form an air force out of the National Liberation Army, with Egyptian, Soviet and Czech assistance. The first aircraft were five ex-Egyptian Air Force Mikoyan MiG-15 fighters which were flown by Egyptian pilots pending the training of Algerians. It is likely that even now there are few Algerian pilots or technicians. Currently the Force Aérienne Algérienne has a personnel strength of some 2,000 men, with 140 Mikoyan MiG-21F interceptors, MiG-17 and MiG-15 fighters in about ten squadrons; two squadrons operating a total of thirty Ilyushin Il-28 light jet bombers; eight Antonov An-12 and four Ilyushin Il-18 transports in one transport squadron; thirty Mil Mi-4 and twenty Sud SA.330 Puma helicopters; twenty-eight Potez Magister armed-trainers; a number of basic training aircraft; and a battalion of SA-2 'Guideline' surface-to-air missiles.

ARGENTINA

Argentinian Air Force *Fuerza Aérea Argentina*

Argentina's history of military aviation began in September 1912 with the founding of a Military Aviation School equipped with

Farman, Blériot and Morane aircraft. Two years later, aircraft were used in manoeuvres by the Argentinian Army, but World War I prevented further development by making new aircraft difficult to obtain. An Italian Aviation Mission in 1919 bought six Ansaldo SVA Primo fighters, four Caproni Ca.33 tri-motor bombers, two Fiat R.2 and two Ansaldo SVA reconnaissance aircraft, plus two Savoia trainers. A French Military Aviation Mission in 1920 reorganized the Military Aviation Service, which by the late 1920s was operating Dewoitine D.21C-1 fighters, Bristol F.2B fighter and Breguet Br.19A-2 reconnaissance aircraft, and Avro 504K trainers.

A Military Aviation Factory was established in 1927, and this was soon licence-building one hundred Avro 504R Gosport trainers, followed by forty each of Bristol F.2B observation aircraft and Dewoitine D.21C-1 fighters.

The following decade saw the MAS organized into six groups (two fighter, two bomber and two reconnaissance) which became three regiments in 1938. A number of nationally-designed aircraft were put into service during this period, including the Aé.M.B.-1 light bomber, the Aé.T-1 liaison aircraft, and the Aé.M.O.-1 trainer and observation aircraft, before the Military Aviation Factory returned to licence production of foreign designs in 1937 with the production of two hundred Curtiss Hawk 75-0 fighters, and five hundred Focke-Wulf Fw 44J Stieglitz trainers. Thirty-five Martin 139-W medium bombers were bought from the United States at this time.

World War II was almost a repeat of World War I, with little progress for the Argentinian armed forces. The exceptions were the supply of a few Douglas DC-2 transports in 1944, and one hundred I.Aé.DL-22 trainers from the former Military Aviation Factory, which had been renamed the Instituto Aerotécnico. In 1944 the Military Aviation Service became the Fuerza Aérea Argentina, a separate service.

Once the war was over, the FAA immediately began to replace its ageing equipment, and during the immediate post-war period new aircraft included one hundred Fiat G.55 fighters, twenty Avro Lancaster and Lincoln bombers, twelve Douglas C-54 and C-47 transports, and thirty Beech AT-11 Kansan crew trainers. These were soon followed by one hundred Gloster Meteor F.4

fighters (which were both Argentina's first jets and the type's first export order), fifty de Havilland Dove, thirty Vickers Valetta and thirty Bristol 170 Freighter transports, two hundred Percival Prentice and thirty Fiat G.46 trainers. There were also one hundred I.Aé-24 Calquin (Eagle) light bombers, and 150 observation-version DL-22s. Argentina became a member of the Organization of American States in 1948.

Little happened for a few years until 1957, when ninety Beech T-34 Mentor trainers were built in Argentina, followed by forty-eight Morane-Saulnier M.S.760 Paris trainers. A couple of years later, a number of North American F-86F Sabre fighter-bombers entered service; and in recent years these have been supplemented by Douglas A-4 Skyhawk fighter-bombers, de Havilland Canada DHC-2 Beaver and DHC-6 Twin Otter transports, and the nationally-designed Dinfia Guarani and Huanquero transport and general-purpose aircraft.

Current equipment of the FAA, which has some 17,000 men, includes one squadron of twelve Dassault Mirage III fighter-bombers, which replaced the last of the Meteors; twenty-five Douglas A-4B Skyhawk and sixteen A-4F Skyhawk fighter-bombers; twenty-five North American F-86F Sabre fighter-bombers; twelve BAC Canberra B.2 bombers; eighty Dinfia Pucana twin-engined turboprop COIN aircraft, which replaced North American T-28s; thirty Douglas C-47 Dakota, and six DC-6, six Lockheed C-130E Hercules, five de Havilland Canada DHC-6 Twin Otter, eight Fokker F-27 Troopship, twenty Hawker Siddeley Dove, fifteen Dinfia Guarani II and thirty five Huanquero transports; a number of light liaison aircraft; six Sikorsky S-55, four Bell UH-1H Iroquois, four 47G Sioux, and fourteen Hughes 269-HM helicopters; and four Dassault Mirage IIIB, thirty-two Morane-Saulnier M.S. 760 Paris, eighty Beech T-34 Mentor, and some North American T-28A trainers.

Argentinian Naval Aviation *Comando de Aviación Naval*

The history of Argentinian Naval Aviation is almost as long as that of the Fuerza Aérea Argentina, dating from 1919 when a Naval Aviation Service was formed with the help of an Italian Mission. Initial equipment consisted of two Macchi M.7 and two

M.9 and two Löhner L-3 flying-boats. During the next few years, additions to this small force included HS.2L and F.5L flying-boats, followed by Dornier Wal and Supermarine Southampton maritime-reconnaissance flying-boats. There were also Vickers Valparaiso general-purpose aircraft, Avro 552 floatplane trainers and Savoia-Marchetti S.59 flying-boat trainers.

Re-equipment in 1937 took the form of Douglas DB-8A-2 attack-bombers, Vought V.65F and V.142 Corsair reconnaissance-bombers, Grumman J2F-2 and Douglas Dolphin amphibians, and Consolidated P2Y-3 flying-boats. The amphibians were embarked aboard cruisers for reconnaissance and spotting duties.

This force remained largely unchanged until 1956, when ten Chance-Vought F4U Corsair fighter-bombers entered service, followed in 1957 by Lockheed P2V-5 Neptune maritime-reconnaissance aircraft, Consolidated PBY-5A Catalina amphibians, and Martin PBM-5 Mariner flying-boats. These were soon joined by Grumman F6F-5 Hellcat fighter-bombers; JRF Goose and J2F-2 amphibians; and North American T-6, Vultee 13T-13 and Beech AT-11 trainers. In 1958 the aircraft carrier HMS *Warrior* was bought from the Royal Navy and named *Independencia*, while in 1969, the Royal Netherlands Navy's *Karel Doorman*, also an ex-RN carrier, was acquired. The *Independencia* was withdrawn from service in 1970.

Current equipment consists of ten Grumman F-9B Panther fighter-bombers; twenty-four Aermacchi MB.326K armed jet trainers – usual carrier-borne equipment; a number of North American T-28 Fennec armed-trainers; six Grumman S-2A Tracker and six Lockheed P-2V5 Neptune maritime-reconnaissance aircraft, with three amphibious Consolidated PBY-5A Catalinas also in this role; five Short Skyvan, de Havilland Canada DHC-6 Twin Otter; fifteen Douglas C-47 and C-54 transports; a number of Grumman TF-9S Cougar and Beech C-45 trainers; some Sikorsky UH-19 and SH-34G, and Bell 47D/G/J Sioux helicopters. Seacat surface-to-air missiles are also employed on some warships, while two Type 42 guided-missile destroyers on order from Great Britain will operate Westland WG.13 Lynx helicopters on anti-submarine duties. A Hawker Siddeley HS 125 was delivered in April 1971.

Argentinian Army Aviation Command

An Air Branch of the Argentinian Army was formed in 1959 to undertake aerial observation post (AOP), liaison and light transport duties. Current equipment includes three each of Douglas C-47 Dakota and de Havilland Canada DHC-6 Twin Otter transports, numbers of Cessna 182J, 310 and Skymaster types, Piper Apache and Aztec aircraft, and seven Bell 206 JetRanger helicopters.

AUSTRALIA

Royal Australian Air Force

Australian involvement in military aviation started in 1913 with the formation of the Australian Flying Corps, which had one squadron with two Royal Aircraft Factory B.E.2a and two Déperdussin aircraft. An AFC B.E.2a and a Farman seaplane were sent with an Australian contingent to German New Guinea in 1914, shortly after the outbreak of World War I, while the following year a small force was attached to the Royal Flying Corps in Mesopotamia fighting against Turkish forces. In 1916 and 1917 four AFC squadrons served in Egypt and England alongside the RFC, which equipped them with Royal Aircraft Factory F.E.2bs and S.E.5as. Although the small force fought with distinction, in 1919 the Australian Flying Corps was disbanded.

The Australian Flying Corps was re-formed in 1920, and the following March became the Australian Air Force, a separate service which gained its 'Royal' prefix some three months later. British assistance was given, largely in the form of the 'Imperial Gift' of war-surplus aircraft to air forces in the Empire. Australia's share of more than a hundred aircraft included D.H.9s, S.E.5as, ten Sopwith fighters, twenty-six Avro 504K trainers and six Fairey IIID seaplanes. In 1925 two squadrons were formed, partly operated by regulars and partly by reservists (Citizen Air Force personnel), for army co-operation duties using the D.H.9s and S.E.5as, while a Fleet Co-operation Flight operated six Supermarine Seagull Mk. III amphibians.

Later during the same decade, Supermarine Southampton flying-boats, Australian-built de Havilland Cirrus Moth trainers, Bristol Bulldog fighters and twenty-eight Westland Wapiti general-purpose biplanes entered service. The overall financial situation prevented further development until 1934, when Hawker Demon fighters, Supermarine Walrus amphibians, and Avro Cadet trainers and Anson reconnaissance aircraft were ordered. After the Ansons had been delivered in 1937, a three-year expansion programme was embarked upon. However, at the outbreak of World War II in 1939, only twelve squadrons out of a planned thirty-two were operational, with 164 aircraft and 3,500 personnel. Equipment consisted of the aircraft already mentioned, plus a squadron of Lockheed Hudson bombers, while nine Short Sunderland flying-boats were given up to the Royal Air Force.

An important development, shortly before the war started, was the formation of the Commonwealth Aircraft Corporation in 1936 by Australian industrialists. The first aircraft produced was the North American NA-33 trainer, known to the Australians as the Wirraway.

At the outbreak of war, Australia offered a six-squadron force for service with the Royal Air Force, but this proved impracticable, although Australians served with RAF squadrons and there were Australian squadrons within the RAF. Altogether, thirteen Australian squadrons were formed within the RAF, and notable amongst these was No. 3 squadron, originally formed as an army co-operation squadron; it later became a fighter squadron and scored the highest number of enemy aircraft shot down of any Allied unit in the Mediterranean.

Although Australian pilots once more distinguished themselves, Australia's main achievement was the organization and operation of the Empire Air Training Scheme, under which the RAAF undertook to train 280 pilots, 184 observer-navigators and wireless-operator/air-gunners a month, plus basic training of a further number who completed their training in Canada. The Wirraway was put into full production, as were Australian-built de Havilland Tiger Moths; and the nationally-designed Commonwealth CA-2 Wackett was developed and two hundred were produced. The United Kingdom provided several hundred Fairey Battles and Avro Ansons.

In 1943, Darwin and several other towns in Australia's Northern Territory were bombed by Japanese aircraft, while the RAAF's equipment in the area only consisted of Lockheed Hudson bombers for retaliatory raids, and Consolidated PBY-5 Catalina amphibians for maritime-reconnaissance. An indigenous fighter, the Commonwealth CA-12 Boomerang, was developed and put into production in double-quick time, proving an effective fighter, but also an excellent ground-attack aircraft. Australian units did in fact serve in every theatre of World War II.

A wide variety of aircraft served with the RAAF during the war, some of which, such as the de Havilland Mosquito, Tiger Moth and Dragon, were built in Australia, along with about four hundred Bristol Beaufighters. Other aircraft included Curtiss A-40 Kittyhawk, Boulton-Paul Defiant, Lockheed Lightning, Hawker Hurricane and Supermarine Spitfire fighters; Bristol Beaufort, Vickers Wellington, Handley Page Hampden and Halifax, Consolidated Liberator, Lockheed Ventura, North American Mitchell and Douglas Boston bombers; a number of former Royal Netherlands Air Force types from Netherlands New Guinea, including six Dornier Do 24, Vultee Vengeance and Curtiss Helldiver dive-bombers; a few flying-boats; Republic P-43 reconnaissance-fighters; and Douglas C-47 Dakota transports. Maximum strength of the RAAF during the war was 200,000 men and women, and there were 10,000 RAAF fatalities.

The end of the war saw the customary reduction in strength, although Australia also contributed to the British Commonwealth Occupation Force in Japan. It was decided that the peacetime complement should be 15,000 men, and that organization should be based on a system of area commands. Re-equipment with more modern aircraft took place during the late 1940s, and included one hundred North American P-51D Mustang fighters produced by Commonwealth Aircraft, seventy-five Avro Lincoln heavy bombers produced by the Government Aircraft Factory, and then eighty Australian-built de Havilland Vampires which replaced the Mustangs. Altogether some 3,000 aircraft were in reserve, and the Citizen Air Force units were re-equipped with Mustangs and Wirraways.

During this period the RAAF was far from being inactive. Douglas C-47 Dakota transports were sent to Europe to help in the Allied airlift to West Berlin, which was cut off from West Germany by the Russians; while an Australian squadron in Japan was attached to the United States Fifth Air Force for duties against the Communist forces in Korea, where it undertook fighter and ground-attack duties using Gloster Meteor F.8s; and a Lincoln and a Dakota squadron operated alongside British forces in Malaya against Communist bandits.

During the early 1950s, North American F-86 Sabres re-equipped the fighter squadrons, while English Electric Canberra jet bombers and de Havilland Vampire jet trainers were also placed in service. All three types were built in Australia, by Commonwealth Aircraft, Government Aircraft Factory, and de Havilland Aircraft respectively. The Australian Sabres had Rolls-Royce Avon engines fitted, which gave them a superior performance compared with the standard aircraft. Modified Avro Lincolns were replaced on maritime-reconnaissance duties by Lockheed P2V-5 Neptunes. In 1955 Australia became a founder-member of the South-East Asia Treaty Organization.

In the 1960s one hundred licence-built Dassault Mirage III fighters, Lockheed C-130 Hercules and de Havilland Canada DHC-4 Caribou transports, and seventy-five licence-built Aermacchi MB.326H all entered RAAF service.

Current strength of the RAAF is 22,500 men. There are four squadrons with one hundred Dassault Mirage III-O fighter-bombers; twenty-four McDonnell Douglas F-4E Phantom IIs have replaced the Canberras (in 1970) and are on lease from the USAF, operating in two or three squadrons pending a decision on whether or not Australia will accept a similar number of General Dynamics F-111 variable-geometry bombers; four F-4E Phantom II tankers are also operated for in-flight refuelling of the combat aircraft; one squadron operates Lockheed P-3B Orion and another operates P-2H Neptune maritime-reconnaissance aircraft; there are eighty-seven Aermacchi MB.326H and sixteen Dassault Mirage IIIB trainers, twenty-four Lockheed C-130A/E Hercules transports and twenty-two de Havilland Canada DHC-4 Caribou transports. Two helicopter squadrons operate Bell UH-1 Iroquois.

Consideration is being given to a replacement, of either New Zealand or Japanese design, for the Commonwealth Winjeel basic trainer. A Mirage squadron is based at RAAF Butterworth in Malaya, while some helicopters have been operated in support of Australian ground forces in South Vietnam. Dassault Mirage F.1 fighters may be ordered as a Mirage III-O replacement.

Royal Australian Navy Fleet Air Arm

The decision to form a Fleet Air Arm for the Royal Australian Navy was taken in 1948; prior to this, amphibious aircraft of the RAAF had been operated from cruisers. Initial equipment consisted of two squadrons each of Hawker Sea Fury fighter-bombers and Fairey Firefly anti-submarine aircraft based on one shore station and on an aircraft carrier, HMAS *Sydney* (formerly HMS *Terrible*), delivered in 1948. A second carrier, HMS *Vengeance* – formerly HMS *Majestic* – was leased in 1953 pending delivery of HMAS *Melbourne* in 1956.

In the meantime, HMAS *Sydney* played an active part in the Korean War, from October 1951. The Sea Furies and Fireflies were replaced by de Havilland Sea Venom F.A.B.53 jet fighters and Fairey Gannet turboprop anti-submarine aircraft respectively, while Bristol Sycamore helicopters and de Havilland Vampire jet trainers also joined the Royal Australian Navy.

After a period as a training carrier, HMAS *Sydney* is now a fast troop carrier, leaving only the extensively modernized HMAS *Melbourne* in service. Currently, the RAN's FAA operates one squadron of ten McDonnell Douglas A-4G Skyhawk fighter-bombers, one squadron of Grumman S-2E Tracker anti-submarine aircraft, and a squadron of some twenty-seven Westland S-58 Wessex anti-submarine helicopters. There are also nine Bell UH-1B Iroquois helicopters. Ten Aermacchi MB.326H jet trainers replaced the Vampires in late 1970 and early 1971.

Australian Army Aviation Corps

Established as a separate Army Corps in 1968, the Australian Army Aviation Corps operates some fifty Bell 47G Sioux, and a number of Sud Alouette III helicopters on AOP and liaison duties. Recently twelve Boeing-Vertol CH-47 Chinook medium-lift helicopters have been placed in service.

AUSTRIA

Austrian Air Force *Österreichische Luftstreitkräfte*

The Austrian Air Force, in common with the Austrian Republic, has had a somewhat chequered history. After the Austrian Republic was formed in 1918 out of the remains of the Austro-Hungarian Empire, the Deutschösterreichische Fliegertruppe (German-Austrian Flying Troop) was formed with former Austrian and German personnel and aircraft of wartime origin. After fighting successfully against Yugoslavia in the Corinthian War of the following year, the Allied Control Commission disbanded the force since the new republic was prohibited from having any military air service.

In 1936, Austria obtained full sovereignty, and plans were laid for the establishment of an air force with six fighter, one bomber, one transport, two observation and a number of training squadrons. A variety of aircraft were acquired, including Fiat C.R.20, C.R.30 and C.R.32 fighters, Junkers Ju 86D bombers and Ju 52/3M transports, Messerschmitt Bf 108B Taifan and Focke-Wulf Fw 58 Weihe communications aircraft, and Fw 44 Stieglitz trainers, the latter being built under licence in Austria. By 1938, when Austria was incorporated into Germany, the combat squadrons were fully operational. During World War II, when the fighter squadrons became the Luftwaffe's Jagdgruppen 1/135 and 1/138, several Austrian pilots distinguished themselves in Europe and North Africa.

The end of the war saw Austria divided into British, French, American and Russian occupied zones, and, for the second time in her short history, she was not allowed to operate military aircraft.

Austria was re-established as a sovereign state in 1955, and immediately the Österreichische Luftstreitkräfte (Olk – Austrian Air Force) was formed with eight Yakovlev Yak-11 and Yak-18 trainers, a gift from the Soviet Union. A small number of

Zlin trainers were purchased soon after, and in 1957 additional trainers, five Fiat G.46-5Bs and three Bell 47G-2 Sioux helicopters, plus six Piper PA-18 Super Cubs and two Cessna 182s for liaison and communications duties. The main event in 1957, however, was the arrival of the Olk's first jets: three de Havilland Vampire trainers. Six each of Westland S-55 Whirlwind and Sud Alouette II helicopters had joined the Olk by 1960.

Austria's first post-war fighter squadron was formed in 1961 with SAAB J-29F Tunnan (Barrel) fighter-bombers, with a second squadron in 1962. Also in 1962, thirty Potez Magisters supplemented the Vampire trainers.

Currently, the Austrian Air Force has some 4,000 men and its aircraft include forty SAAB 105E strike aircraft in two squadrons; these replaced the J-29Fs, a transport squadron operating six de Havilland Canada DHC-2 Beavers and two Short Skyvan 3Ms, two Sikorsky S-65 helicopters, fourteen Sud Alouette III and nine Alouette II helicopters, and twenty-two Agusta-Bell 204B and twelve JetRanger helicopters. Twenty-one SAAB Safirs operate in the training role, along with an assortment of de Havilland Vampires, Potez Magisters and North American T-6Gs.

BELGIUM

Belgian Air Force *Force Aérienne Belge*

The origins of the Force Aérienne Belge can be traced back to the formation of an army flying corps. Farman H.F.3 biplanes were manufactured under licence in Belgium, and during 1912 a Belgian Army Farman F.20 was used for the first European trials of the new American Lewis gun. During this period the flying corps had merely been a section of the Army Balloon Company, but in 1913 it became the Compagnie des Aviateurs, with sixteen aircraft in four reconnaissance squadrons.

A number of privately-owned aircraft were pressed into service in 1914 at the start of World War I. In 1915 the Compagnie des Aviateurs had its name changed to the Aviation Militaire, but in spite of a reorganization the Belgian war effort was seriously hampered by an acute shortage of pilots and the lack of a suitable training organization. Nevertheless, although many such pilots were trained in Great Britain only at their own expense, there were sufficient to form a naval air squadron with Short 225D seaplanes for service in Belgian Central Africa. Wartime aircraft included Royal Aircraft Factory B.E.2as, R.E.8s, Sopwith 1½-Strutters, Pups and Camels, D.H.9s, Farmans, Spads, Hanriot H.D.1s, Caudron G.IIIs and IVs, and Breguet Br.14s.

After the war, the usual run-down in strength was accompanied by a further reorganization. The Aviation Militaire was formed into eight groups with a total of twenty-six squadrons: 1st Group operated balloons; 2nd (Observation) Group operated Ansaldo A.300 and D.H.4s; 3rd (Army Co-operation) Group operated D.H.4s and F.2Bs; 4th and 5th (Fighter) Groups operated Nieuport 27C and Spad 13 fighters; 6th (Reconnaissance) Group operated D.H.4s and D.H.9s, and Breguet Br.14s; 7th (Technical) Group and 8th (Flying School) Group operated an assortment of aircraft, including Avro 504Ks and Fokker DVIIs. Some of these aircraft were licence-built in Belgium.

Further reorganization followed in 1925, with the then twenty-six squadrons formed into three wings, which became regiments in 1927. New aircraft types introduced during the late 1920s included Avia BH-21 fighters and Breguet Br.19 reconnaissance-bombers, plus a number of trainers, including Stampe et Vertongen RSV 32/100s and RS 26/180s, and Morane-Saulnier M.S.230s. The scope for domestic production of aircraft extended in 1931 when Britain's Fairey Aviation opened a Belgian branch factory. This produced Fairey Fireflies and Foxes for the fighter squadrons, while Avro 504N and 626 Prefects entered service, along with the Belgian Renard R.31 reconnaissance aircraft. During the latter part of the 1930s, Gloster Gladiator fighters, Fokker F.VII tri-motor transports and Stampe SU-5 general-purpose biplanes entered service, along with Koolhaven F.K.56 trainers.

The outbreak of World War II in 1939 saw Belgian production of Fiat C.R.42 and Hawker Hurricane fighters, and Fairey Battle light bombers. A few British Hurricanes and some Airspeed Oxford crew-trainers were supplied before the outbreak of war, but most of Aviation Militaire's aircraft were destroyed on the ground on 10th May 1940, when Germany invaded. The few which did manage to get into the air fought valiantly against overwhelming odds. However, the flying school's squadrons escaped to French Morocco.

A number of Belgian personnel reached the United Kingdom from Europe and North Africa and were absorbed into the Royal Air Force as members of Belgian flights attached to RAF squadrons. Aviation Militaire personnel serving in the Belgian Congo joined the South African Air Force, and fought in North Africa. Later two Belgian fighter squadrons were formed within the RAF, with a combined score of 161 confirmed victories, 37 probables and 61 damaged enemy aircraft. Some 1,200 Belgian personnel served with the RAF during the war.

It was in 1946, on the disbanding of the Belgian element of the RAF, that the Force Aérienne Belge was formed, the plan being that it should be a separate service with an initial strength of four day- and one night-fighter wings, with a transport and an army co-operation wing. Equipment initially included Supermarine Spitfire day- and de Havilland Mosquito night-fighters, Douglas C-47 Dakota, Avro Anson, Airspeed Oxford and de Havilland Dominie transport and communications aircraft, and Percival Proctors for army co-operation duties. Belgium's first jets, Gloster Meteors, replaced the Spitfires in 1949, and these were supplemented by Republic F-84 Thunderjets in 1951, marking the start of American military aid when Belgium joined the North Atlantic Treaty Organization. The Mosquitoes were replaced by Meteors, which were in turn replaced by Avro Canada CF-100 all-weather fighters. Other aircraft operated by the FAB during the immediate post-war period included Douglas C-54 and Fairchild C-119F Packets (for transport services to the Belgian Congo), Bristol Sycamore helicopters and Hunting Pembroke light transports. A number of Meteor and Lockheed T-33A jet trainers were also placed in service, but prior to 1955, a proportion of pilots received their training in the United States. An interesting feature of the FAB's pilot training at this time was the posting of advanced students to the Congo to take advantage of the better flying conditions.

During the latter part of the 1950s, the FAB received licence-produced Hawker Hunter fighters, and, from the United States, Republic F-84F Thunderstreak fighter-bombers and RF-84F Thunderflash reconnaissance aircraft to replace the earlier F-84s.

Currently the FAB has 20,500 men, and operates two Lockheed F-104G interceptor squadrons with about fifty aircraft, plus a similar number of these aircraft in another two squadrons for strike duties; there are sixty-three Dassault Mirage 5-BRs in two reconnaissance squadrons, and twenty-seven Mirage 5-BAs in a fighter-bomber squadron – the Mirages replaced the F-84Fs and RF-84Fs. Thirty-three Fairchild C-119 Packet transports are likely to be replaced by Franco-German Transalls, but another twenty-three transport aircraft include Douglas C-47 Dakota, C-54, and DC-6 and Hunting Pembroke types. Training is on thirty-six SIAI-Marchetti SF.260s, a number of Lockheed T-33As and forty Potez Magisters, plus seventeen Dassault Mirage 5-BDs. Five each of Sikorsky S-58C and Sud H-34A (licence-built S-58s) helicopters are operated. There are eight Nike-Hercules surface-to-air missile squadrons. Twelve Lockheed C-130H Hercules transports are on order for 1972–3 delivery.

Belgian Navy   *Force Navale Belge*

The Belgian Navy operates two Sikorsky S-58C and three Sud Alouette III helicopters, the latter from suitably equipped warships.

Belgian Army Aviation   *Force Terrestre Belge*

The Belgian Army operates four squadrons of light aircraft and helicopters on liaison, air observation-post and allied duties. This force has existed since the end of the war, initially operating Auster AOP 6s, but at present twelve Dornier Do 27s form the fixed-wing element, while there are also thirty-eight Sud Alouette II and forty-two Sud Alouette III helicopters.

BOLIVIA

Bolivian Air Force   *Fuerza Aérea Boliviana*

Although Bolivian Army officers were given flying training in 1917, it was not until 1924 that an air arm, the Cuerpo de

Aviación, was formed. A variety of aircraft were operated during the first few years, including Breguet Br.19A-2 and Fokker C.V-C reconnaissance-bombers, with Caudron C.97 and Morane-Saulnier M.S. 139 trainers. In 1929, Vickers Vespa III AOP biplanes and 143 fighters were obtained, and soon followed by Curtiss Hawk IA fighters, Junkers W.34 bombers, Curtiss-Wright Osprey general-purpose aircraft, and Curtiss Falcon AOP aircraft. All these types, totalling some sixty aircraft, were used during a war with the neighbouring state of Paraguay during the latter part of the 1920s and the early 1930s. Paraguay won this war.

An Italian Air Mission in 1937 attempted to reorganize the force. In 1940, Curtiss-Wright 19R and CW-22 trainers entered service and, the following year, three Junkers Ju 86 transports were expropriated from a German airline operating within Bolivia. Also during 1941, an American Military Aviation Mission reorganized the force by dividing Bolivia into four air defence zones, and it was at this time that the present title, Fuerza Aérea Boliviana, was adopted, although the FAB was still a part of the Bolivian Army. New aircraft placed in service at this time included Grumman OA-9 amphibians, Douglas C-47 Dakota transports, Stinson 105 Voyager and Interstate L-8 AOP aircraft, plus an assortment of trainers including Beech AT-11 and North American NA-16-3 aircraft.

American entry into World War II cut Bolivia off from new aircraft supplies, but as a founder-member of the Organization of American States in 1948, Bolivia became eligible for American military aid. Initially this included Republic F-47D Thunderbolt fighter-bombers, North American B-25J Mitchell bombers, and T-6 Texan trainers. In 1958 eight Boeing B-17G Fortress bombers were also received from the United States, and converted into transports.

In recent years the FAB has been equipped by the United States with small numbers of aircraft suitable for counter-insurgency operations (COIN). Current equipment includes one squadron with twelve North American F-51D Mustang fighter-bombers, and three AT-6 armed-trainers; a further twenty North American T-6 Texan and T-28 trainers; with seven Cessna 185 liaison aircraft; eighteen Douglas C-47 Dakota and a C-54 transport aircraft; and twelve Hughes 500M and three Hiller H-23 helicopters.

BRAZIL

Brazilian Air Force   *Forca Aérea Brasileira*

Although the first officially recognized flight in Europe in 1906 was by a Brazilian, it was not until 1913 that a Brazilian Navy seaplane school was formed with an Italian Bossi seaplane. The Brazilian Army followed this example soon afterwards, and in the following year a combined total of seventeen aircraft were operated, including Farmans and Blériots. In 1917, Brazilian officers were sent to the United Kingdom for flying training, while throughout World War I, the UK provided Brazil with over eighty aircraft.

After the war ended, the Army received French assistance in the formation of a Brazilian Army Air Service which in 1922 was operating one squadron of Breguet Br.14A-2 and one squadron of Spad S.7 reconnaissance aircraft, while a variety of British and French aircraft were employed on training duties. The Brazilian Naval Air Service had by this time reached a three-squadron strength, operating fourteen Savoia reconnaissance-seaplanes, twelve F.5L flying-boats, and twelve each of Avro 504 and Curtiss trainers.

Aircraft operated during the inter-war period by the Army included Boeing 256 and 267 and Curtiss Hawk 75A fighters; North American NA-44 and Vultee V-119B light bombers; Vought O2U-1 Corsair, de Havilland Fox Moth, Gipsy Moth and Tiger Moth, Beech D-17A, Morane-Saulnier M.S.230 and Avro 630 aircraft on training and communications duties. The Navy operated Boeing F4B-4 fighters, Curtiss-Wright Osprey general-purpose aircraft, and Fairey Gordon reconnaissance seaplanes. Both services received a number of Brazilian-designed aircraft, including the Muniz M-7 and M-9 trainers, and Brazilian-built Focke-Wulf Fw 44J Stieglitz and Fw 58B Weihe trainers. Apart from the normal military duties, the Army also undertook aerial-survey work and the provision of mail services.

Brazil declared war on Germany and Italy in 1942, after Brazilian ships had been torpedoed by German U-boats. Brazilian bases were put at the disposal of the Allies, and military assistance was received in return from the United States. Two years earlier the Brazilian Army and Navy Air Services had merged to form a separate service, the Forca Aérea Brasileira, and the new air force benefited greatly from the delivery of one hundred Curtiss P-40 Warhawk fighters; a handful of Douglas B-18B and Lockheed A-28 bombers; twenty-five North American B-25 Mitchell bombers; one hundred Vultee BT-15 Valiant, two hundred licence-built Fairchild PT-19, 125 North American AT-6, ten Beech AT-7 and ten AT-11 trainers. Further supplies came as war progressed, and in 1944 a Brazilian squadron was equipped with Republic F-47D Thunderbolt fighter-bombers and sent to Italy to fight against the Axis forces there. In 1944 and 1945 were acquired additional Thunderbolts and twenty-nine Vultee A-35B Vengeance dive-bombers; twenty-five each of Douglas A-20 and North American B-25 Mitchell bombers; twenty-one Piper L-4 and forty Aeronca L-3 liaison aircraft; eight Beech C-45, eleven Douglas C-47 Dakota, eight Lockheed C-60 and thirty-three Cessna UC-78 transports. Brazil had about 1,000 military aircraft when the war ended, although the numbers then declined rapidly.

Brazil was a founder-member of the Organization of American States in 1948, and once again received a considerable quantity of American military aid, including additional Thunderbolt and Mitchell aircraft, some Boeing B-17G Fortress and Lockheed PV-2 Neptune maritime-reconnaissance aircraft, and North American T-6 Texan trainers. Brazil's first jets, sixty Gloster Meteor F.8 fighters and T.7 trainers, were delivered in 1954, while in 1956 twelve Fairchild C-82 transports were delivered and licence-production of the Fokker S.11 Instructor trainer started.

Additional jet equipment, in the form of fifty Lockheed F-80C Shooting Stars and a similar number of T-33A jet trainers, arrived during the second half of the 1950s, along with Fairchild C-119 Packet transports. These were followed by ten Lockheed C-130E Hercules transports, a Vickers Viscount, various Beech light transports, and six Hawker Siddeley HS 748 transports. In 1965 all fixed-wing naval aircraft, including those operating from the aircraft carrier *Minas Gerais* (formerly HMS *Vengeance* before her acquisition in 1957), passed to the FAB. The naval aircraft included a number of Grumman S-2A Trackers.

Currently the Forca Aérea Brasileira, with some 30,000 men, operates sixteen Dassault Mirage III and fifteen Douglas A-4F Skyhawk fighter-bombers; eighteen Douglas B-26K bombers and fifty-four Lockheed TF-33 armed jet trainers; fourteen Lockheed PV-2 Neptune maritime-reconnaissance aircraft; thirteen Grumman S-2A Tracker anti-submarine aircraft, and twelve Grumman HU-16 Albatross amphibians on naval support duties; twenty-four de Havilland DHC-5 Buffalo, ten Lockheed C-130E Hercules, five Douglas C-54 and forty C-47 Dakota, twelve Fairchild C-119 Packet, and numbers of Beech C-45, Hawker Siddeley HS 748 and HS 125, a BAC One-Eleven, and some other transports; six Bell SH-1D, six UH-1D Iroquois, seven JetRanger, and eighteen 47G/J, and five Sikorsky UH-19D helicopters; 150 Neiva IPD-6201 Universal, seven Potez Magister, and a number of other training types including North American T-6G Texan, T-28, Beech and Fokker aircraft, much of which are being replaced by 112 Aermacchi MB.326 trainers being built in Brazil.

Naval Air Arm *Forca Aeronavale*

The Forca Aeronavale was formed as the Marinha in the early 1950s, after the Brazilian Navy had received considerable American assistance. Initial equipment included three Bell 47J Sioux helicopters and these were soon joined by two Westland Wigeon helicopters. A former Royal Navy aircraft carrier, HMS *Vengeance*, was acquired in 1957 and renamed the *Minas Gerais*, from which were operated thirteen Grumman S-2A Tracker antisubmarine aircraft and some North American T-28C armed trainers. In 1965 the fixed-wing aircraft passed to the Forca Aérea Brasileira, although carrier-borne operations continue.

Currently, the Forca Aeronavale operates a helicopter fleet of fifteen Westland Whirlwinds, five Wasps and two Wigeons, two Bell 47J Sioux, three Hughes 269A, nine Hughes 200, and six Hughes 500, and four Sikorsky SH-3D Sea Kings.

BULGARIA

Bulgarian Air Force

The history of Bulgarian air power dates from the Balkan War of 1912-13, when an Army Aviation Corps operated briefly against the Turks using a dozen Blériot and Bristol monoplanes flown by foreign pilots. Although this force was disbanded after the war ended, it was revived in 1915 with German and Austrian assistance after Bulgaria had allied herself with the Central Powers. Germany and Austria supplied both aircraft and pilots. After the Allied victory in 1918 the Bulgarian Army Aviation Corps was disbanded once again, and under the terms of the Treaty of Neuilly a Bulgarian military air arm was forbidden.

In 1937, Bulgaria denounced the Treaty and established the Bulgarian Air Force as an integral part of the Bulgarian Army. The first aircraft were Polish, including twenty-four PZL P-24-G fighters, forty-three PZL P-43 reconnaissance-bombers, Avia B.534 fighter and Letov S.328 reconnaissance-biplanes; and there were Bulgarian-built Focke-Wulf Fw 44 Stieglitz trainers. Bulgaria joined the Axis powers in 1941, and German forces were able to use Bulgarian bases for the invasion of Greece. Bulgaria also received German military aid, including twenty Messerschmitt Bf 109E fighters, Junkers Ju 86D and Ju 87B, and Dornier Do 17M bombers, which were followed by Focke-Wulf Fw 58 Weihe communications aircraft and Arado Ar 96 trainers. Luftwaffe instructors and advisers were seconded to the Bulgarian Air Force. The largest deliveries of aircraft to the Bulgarians occurred in 1943 when one hundred Dewoitine D.520 and 150 Messerschmitt Bf 109E fighters were supplied to enable the Bulgarian Air Force to replace Luftwaffe units posted to the Russian front.

Although Bulgaria never declared war on the Soviet Union, and indeed often pleaded neutrality, the Soviet Union invaded Bulgaria towards the end of 1944. Bulgaria was limited to a post-war air force strength of 5,000 men and ninety aircraft, but by 1950 a policy of expansion was well under way with Soviet assistance. Initially this included Yakovlev Yak-9 fighters and Ilyushin Il-2 close-support aircraft, with a few bomber, transport and training types, the latter mainly consisting of Yak-18s. The first jet equipment arrived in 1953: twenty-four Yak-23 fighters followed shortly afterwards by some sixty Mikoyan MiG-15s, and the obsolescent Ilyushin Il-10 ground-attack aircraft.

Currently the Bulgarian Air Force has 12,000 men and, as a member of the Warsaw Pact, continues to receive Soviet military aid. Two interceptor squadrons operate twenty-four Mikoyan MiG-21s, while six squadrons operate seventy-two MiG-19 fighters; eleven squadrons operate some 130 MiG-17s, mainly on ground-attack duties; and there are also twenty-four MiG-17Cs in two reconnaissance squadrons, while a third squadron operates Ilyushin Il-28 reconnaissance-bombers. Altogether some thirty transports are operated, mainly Ilyushin Il-14s, but with a few Il-12s and possibly one or two Lisunov Li-2s (Douglas C-47 built in Russia). There are thirty Mil Mi-4 helicopters. Training is on MiG-15UTI, L-29 Delfin, and Yak-11 and Yak-18 aircraft.

BURMA

Union of Burma Air Force

The Union of Burma Air Force was founded with British assistance in 1955, and initial equipment consisted of ex-RAF Supermarine Spitfire Mk. 18 fighters, de Havilland Mosquito Mk. 6 fighter-bombers, and Airspeed Oxford and de Havilland Tiger Moth trainers. These were soon supplemented by Spitfire Mk. 9s, followed by Hawker Sea Fury F.B.11 fighter-bombers, Bristol 170 Mk. 31M Freighters, Douglas C-47 Dakotas and Beech D18Ss which replaced the earlier fighter-bombers and transports. A number of Cessna 180 liaison aircraft were also received. In each case, the aircraft concerned were delivered in very small batches, no fairly large orders arriving until the Hunting Provost T.53 trainers of the late 1950s. A few de Havilland Vampire T.55 jet trainers were also received, and, in 1960 de Havilland Canada DHC-3 Otters were first introduced. Japan provided a few Kawasaki-Bell 47 Sioux helicopters, also in 1960.

Currently the Union of Burma Air Force has some 6,000 men, and operates one squadron with twelve North American F-86F Sabre fighter-bombers; one squadron with ten Lockheed T-33A and six de Havilland Vampire T.55 armed jet trainers; four Beech C-45, six Douglas C-47 Dakota, six de Havilland Canada DHC-3 Otter and two Bristol 170 Mk. 31M Freighter transports; six Kawasaki-Bell 47 Sioux, ten Kaman Huskie, eight Sud Alouette III and three Mil Mi-4 helicopters; with some thirty

Hunting Provost T.53, and ten de Havilland Canada DHC-1 Chipmunk trainers, and ten Cessna 180 liaison aircraft.

CAMBODIA

Cambodian Air Force

Formerly part of French Indo-China, Cambodia became an independent kingdom within the French Union in 1949, and gained complete independence in 1955. In 1953 plans were laid by the Government for the formation of an air force bearing the title of Royal Khmer Aviation. Initial equipment consisted of seven Fletcher FD-25A/13 Defender light attack aircraft for police duties, followed in 1955 by seven Morane-Saulnier M.S.733 Alcyon trainers. The United States started to provide military aid in 1956 with eight Cessna L-19 Bird Dog AOP aircraft, and a handful of Douglas C-47 Dakota and de Havilland Canada DHC-2 Beaver transports.

During the next few years, thirty Douglas A-1D Skyraiders were provided from French surplus stocks, and other aircraft were received from both sides of the Iron Curtain. These included Curtiss C-46 Commando transports, Potez Magister and North American T-28D armed-trainers, plus Cessna T-37B and North American T-6G trainers, Sikorsky S-58 and Sud Alouette II helicopters from the West, and Mikoyan MiG-15 and MiG-17 fighter-bombers and Antonov An-2 transports from the East.

In 1970 a change of government led to the overthrow of a 'neutralist' regime which had permitted North Vietnamese troops to use Cambodian territory as a base for operations in South Vietnam. A pro-Western government made Cambodia a republic, and the title of the Cambodian Air Force was adopted. The Mikoyan MiG-15 and MiG-17 fighter-bombers were destroyed by Viet Cong action early in 1971. Future supplies are now likely to be entirely from Western sources, and probably include additional North American T-28s.

Currently the Cambodian Air Force has some 2,500 men, with one squadron each of twenty Douglas A-1 Skyraider and North American T-28 Trojan ground-attack aircraft; twelve

Douglas C-47 Dakota, two de Havilland Canada DHC-2 Beaver, six Dassault M.D.315 Flamant, and a handful of Antonov An-2 and Ilyushin Il-14 transports; ten Sud Alouette II, three Sikorsky S-58, and a Mil Mi-4 helicopters; with four Potez Magister, twenty Sud Horizon, six North American T-6G Texan and six Cessna T-37A trainers; and eight Cessna L-19 Bird Dog AOP aircraft.

CANADA

Canadian Armed Forces   *Forces Armées Canadiennes*

The history of Canadian military aviation dates from the formation of a Canadian Aviation Corps in 1914, with one aircraft which was scrapped the following year although it had accompanied a Canadian Army contingent to Europe on the outbreak of World War I. Many Canadians flew with the Royal Flying Corps, which maintained its own flying school in Canada. Canadian production of Avro 504 and Curtiss JN-4 aircraft started in 1916, while two years later two Canadian squadrons were formed to operate S.E.5A fighters and D.H.9A bombers. It was also in 1918 that the Royal Canadian Naval Air Service was founded, to be disbanded in 1919.

The Canadian Air Force was formed in 1920 under Army control, with British assistance including a share of the 'Imperial Gift' of war-surplus aircraft for the Empire. Canada received eighty aircraft of Airco, Avro, Bristol, Curtiss and Sopwith design. In 1923, King George V bestowed the 'Royal' prefix on the new air force. Part of the RCAF's initial strength included Curtiss H.16 flying-boats, which were replaced in 1924 by eight Vickers Vimy flying-boats, and supplemented the following year by Canadian-built Vickers Vedette amphibians.

Most of the RCAF's duties during the 1920s were of a civil nature, including survey and some types of transport and communications work in the less densely populated areas. New aircraft entered service, including Armstrong-Whitworth Siskin III fighters and Atlas liaison and AOP aircraft, Ford 6-AT Trimotor transports, de Havilland Gipsy Moth, Curtiss-Reid

Rambler, Avro Avian and Tutor, and Fleet Trainers. A few Fairchild 71-C seaplanes were introduced before the overall economic situation caused a 20 per cent cut in strength in 1932. However, an increase in Non-Permanent (reserve) Air Force strength was made to compensate for the cuts in the regular service.

An expansion programme began in 1935. In the following year, most civil duties were transferred to the Department of Transport. A number of new squadrons were formed with obsolete Westland Wapitis, ex-RAF. A few Vickers Vancouver flying-boats entered service, followed by eighteen Supermarine Stranraer maritime-reconnaissance flying-boats and twenty Northrop Delta transports. The worsening international situation acted as a spur to RCAF re-equipment and expansion, with a fair proportion of the aircraft being built in Canada. During this period the RCAF received Grumman GE-23 fighters, Bristol Blenheim bombers, Blackburn Shark torpedo-bombers, Noorduyn Norseman transports and North American NA-16-3 trainers.

There were other developments at this time, including the formation of the RCAF into three commands, Eastern, Western and Training, but most notable was the elevation in 1938 of the RCAF from being under Army control to being a separate service.

A British Air Mission visited Canada in 1939 and, apart from Canadian construction of aircraft for the RAF, one of the results of their visit was Canadian participation in the Empire Air Training Scheme, in which Canada trained pilots, many of whom had had their basic training in Australia, for the RAF and other air forces. While Canadian production of military aircraft got under way, the RCAF received a few Hawker Hurricane fighters and Fairey Battle bombers, along with some Westland Lysander army co-operation aircraft.

The outbreak of World War II found the RCAF operating 270 aircraft, of which only twenty were Hawker Hurricane fighters, ten Fairey Battle bombers and another ten Blackburn Shark torpedo-bombers, while the rest were much more elderly aircraft including eight Supermarine Stranraer and four Vickers Vancouver flying-boats, a few Westland Wapiti bomber conversions, and most of the other post-World War I aircraft already mentioned. The Airspeed Oxford and the Avro Anson trainers were two new second-line acquisitions. There were eight regular and twelve reserve squadrons. A Hurricane squadron played an important part in the Battle of Britain, and by 1941 no less than sixteen RCAF squadrons were operating from the United Kingdom.

Aircraft operated by the RCAF during the war included Supermarine Spitfire, Hawker Hurricane and Typhoon, Curtiss Kittyhawk and Tomahawk, Bell Airacobra, de Havilland Mosquito, Bristol Beaufighter and North American Mustang fighters, night-fighters and fighter-bombers; Fairey Battle light bombers and Albacore torpedo-bombers; Vickers Wellington, Handley Page Hampden and Halifax, Avro Manchester, Lancaster and Lincoln, Bristol Blenheim, Boulton-Paul Defiant and Consolidated Liberator bombers; Northrop Nomad, Lockheed Hudson and Ventura, and Douglas Digby maritime-reconnaissance bombers; Consolidated PBY-5 Catalina and Boeing PBY-5A amphibians and Saunders-Roe Lerwick flying-boats were also employed on maritime-reconnaissance duties; Lockheed Lodestar and Douglas C-47 Dakota transports; de Havilland Tiger Moth, Fleet Fort and Finch, Fairchild Cornell, Boeing-Stearman PT-27 and Cessna T-50 trainers. That many of these were Canadian-built gives an indication of the capability of Canadian industry, which was also building aircraft for the RAF!

For the most part the RCAF operated in Europe and over the North Atlantic, but some squadrons were deployed in the Pacific theatre, while in 1942 a unit was sent to Ceylon. No less than 8,000 RCAF personnel were decorated by Allied and Commonwealth governments, while more than 17,000 died in action: totals which were second only to those of the RAF.

Post-war duties of the RCAF included being a part of the occupation forces in Germany. Strength was cut from the wartime peak of ninety squadrons to eight regular and fifteen reserve units, which were to be a base for a limited post-war re-expansion.

The first post-war aircraft were North American B-25 Mitchell light bombers and Beech C-45F Expeditor light transports, along with additional Mustangs and Lodestars. A long-range transport element, based on Canadair DC-4M North Stars (Canadian-built Douglas C-54), was formed at this time, and the first jets, de Havilland Vampire F.8 fighters, were ordered, deliveries taking place in 1948. Additional jet equipment followed, including the Avro Canada CF-100 Canuck all-weather fighter and Canadair F-86E Sabre fighter-bombers (licence-built North American Sabres), with licence-built Lockheed T-33A trainers. During the early 1950s twelve Sabre and CF-100 squadrons were based in Germany. Canada was a founder-member of the North Atlantic Treaty Organization in 1949.

The early post-war period saw Avro Lancaster and Lincoln heavy bombers employed on maritime-reconnaissance duties, but these were replaced by Lockheed P2V-7 Neptunes, which in turn were replaced by Canadair Argus aircraft, a piston-engined maritime-reconnaissance development of the Canadair CL-44 (civil) or Yukon (military) long-range transports which were in fact Canadian versions of the Bristol Britannia turboprop airliner. Some fifty Fairchild C-119F Packet transports entered Canadian service, accompanied by Canadair CC-109 Cosmopolitans (licence-built turboprop Convair Metropolitans) which replaced most of the wartime Dakotas, and later supplemented by the Yukons, which had Rolls-Royce Tyne turboprops in place of the Bristol Siddeley engines of the Britannia. Two de Havilland Comet 1A jet airliners were also bought, but later converted for radar calibration duties – these were eventually the last Comet 1s to be withdrawn. Other aircraft of the period included de Havilland Canada DHC-1 Chipmunk trainers, Vertol H-21A, Sikorsky S-51, S-55 (or H-19) and S-58, and Bell 47G Sioux helicopters. De Havilland Canada, the world's largest manufacturer of short take-off (STOL) transports provided the RCAF with DHC-2 Beavers, DHC-3 Otters, DHC-4 Caribou, and DHC-5 Buffalo transports during the late 1950s and throughout the 1960s.

Canada provided air support for the United Nations, initially in the Suez Canal zone, and after 1956 on a number of occasions – usually by providing STOL transport aircraft.

Canadian-built Lockheed CF-104 Starfighters replaced the Sabres in Germany during the 1960s; initially there were six interceptor-strike squadrons and two reconnaissance squadrons, but towards the end of the decade this force was reduced to two squadrons. Canadair-built Northrop CF-5A Freedom Fighters were introduced during 1969 and 1970 and may be operated in Germany, although at present it seems as if only twenty of the seventy-one CF-5As and a few of the forty-six CF-5B trainers will see service, in Canadian colours anyway. Fifty-eight Canadian-built McDonnell CF-101 Voodoo interceptors replaced the CF-100s during the early 1960s, only to be replaced in 1969 by sixty-six US-built F-101s which have a superior performance.

One major post-war development was the formation of an air arm for the Royal Canadian Navy following a reorganization. Four Royal Navy squadrons manned by Canadian personnel were transferred to Canada, and a carrier, HMCS *Warrior* obtained from the United Kingdom in 1946, operated two of these squadrons, one with Supermarine Seafire fighters and the other with Fairey Firefly anti-submarine aircraft. In 1948, HMCS *Warrior* went into reserve and HMCS *Magnificent* entered service with Hawker Sea Fury fighters, and about one hundred Grumman TBM-3E Avengers were obtained from the United States; most of these anti-submarine aircraft were shore-based.

HMCS *Magnificent* was only on loan from the Royal Navy, and in 1957 the former HMS *Powerful* was commissioned as HMCS *Bonaventure*. Until her withdrawal from service in 1968, this ship operated McDonnell F2H-3 Banshee fighters and Grumman S2F-1 Tracker anti-submarine aircraft. Sikorsky S.55 and Bell HTL-4 helicopters were also operated by the Royal Canadian Navy from the aircraft carriers and from some other naval vessels.

A Canadian Army AOP unit was formed in 1946 with Auster aircraft, which were replaced in 1954 by Cessna L-19A Bird Dogs. Bell 47 Sioux and Sikorsky S-51 helicopters were also operated.

In 1967, Canada's centenary year, the Canadian Army, Royal Canadian Navy and Royal Canadian Air Force were formed into a unified defence force: the Canadian Armed Forces. All service ranks, which hitherto had been modelled on RAF and Royal Navy lines, became the equivalent now in what had been hitherto army ranks. The CAF, as it is called, is divided up into a number of commands, in addition to Canadian Armed Forces Headquarters; these are Training Command, Material Command, Mobile Command, Maritime Command, Air Defence Command, Air Transport Command, and Canadian Armed Forces in Germany.

Current strength of the CAF is 90,000, of which 40,000 are engaged in former RCAF units and activities. Currently, Air

Defence Command operates three McDonnell Douglas F-101 Voodoo interceptor squadrons with sixty-six aircraft, and a number of Bomarc B surface-to-air missile squadrons. Mobile Command operates two Canadair CF-5A fighter-bomber squadrons with about twenty-six aircraft, fifty Bell CUH-1N Iroquois and seventy-four OH-58A Kiowa helicopters, and fifteen de Havilland Canada DHC-5 Buffalo STOL transports. Air Transport Command operates five Boeing 707-320C transports which can be easily converted into tankers for in-flight refuelling, two squadrons of Lockheed C-130E Hercules (of which there are twenty-three), a squadron of Canadair CC-106 Yukon transports and a squadron of Canadair Cosmopolitan transports, while a number of DHC-4 Caribou and Douglas C-47 Dakotas remain, and six squadrons of DHC-3 Otters are in reserve. Canadian Armed Forces in Germany also operate some of the above on rotation, plus forty Lockheed CF-104 Starfighters in two reconnaissance squadrons. Maritime Command – the former Royal Canadian Navy and also the RCAF's maritime-reconnaissance squadrons – operates four long-range maritime-reconnaissance squadrons which are currently replacing their Canadair CL-28 Argus aircraft with thirty licence-built Lockheed P-3B Orions, a squadron of Grumman S2F-3 Tracker anti-submarine aircraft, and a squadron of Sikorsky CHSS-3 Sea King anti-submarine helicopters which usually operate singly from a fleet of nine destroyers capable of operating helicopters. Training Command operates a fleet of Beech Musketeers, which are currently replacing Chipmunks, Canadair CL-4 Tutors, a few Canadair CF-5Bs, Beech C-45 Expeditor and Douglas C-47 Dakota aircraft, and Hiller UH-12 helicopters. Eight de Havilland Canada DHC-6 Twin Otters were ordered in 1971 for rescue duties.

CENTRAL AFRICAN REPUBLIC

Central African Air Force   *Force Aérienne Centrafricaine*

The Central African Republic gained independence from France in 1960, and has remained in the French Community since independence, receiving military aid from France. A small air force exists using the standard French aid package of a Douglas C-47 Dakota, three Max Holste 1521M Broussards (Bushrangers) and a Sud Alouette II helicopter. Some aid has been recently provided on a small scale by the United States.

CEYLON

Royal Ceylon Air Force

The Royal Ceylon Air Force dates from 1950, when it was formed with RAF assistance. After some initial uncertainty, twelve de Havilland Canada DHC-1 Chipmunk and nine Boulton-Paul Balliol trainers were purchased in 1953, and a couple of Airspeed Oxford training and communications aircraft followed. In 1955 the first de Havilland Dove light transport arrived, and other aircraft received at this time included Scottish Aviation Pioneers and Westland Dragonfly helicopters. Eight Hunting (now BAC) Jet Provost armed-trainers were delivered in 1958, with additional Doves and Pioneers.

Operations have been directed against smuggling and evasion of immigration controls, plus the internal security duties common in the area.

Current equipment includes eight BAC Jet Provost armed trainers, nine de Havilland Canada Chipmunks, five de Havilland Devon and four Heron light transports, with three Scottish Aviation Pioneers also engaged in this role, and a Bell Jet-Ranger, a Hiller UH-12C and two Westland Dragonfly helicopters. Russian aid was provided early in 1971 to help quell an armed rebellion; this included six MiG-17 fighters and Soviet personnel.

CHAD REPUBLIC

Chad Air Squadron *Escadrille Tchadienne*

Chad gained independence from France in 1960 and, in common with other members of the French Community, this former region of French Territorial Africa received military aid. The standard package of a Douglas C-47 Dakota, three Max Holste 1521M Broussards and a Sud Alouette II helicopter is operated.

CHILE

Chilean Air Force *Fuerza Aérea de Chile*

Formation of a flying school in 1913 marked the start of Chilean military aviation, and this was soon followed by the creation of a Chilean Military Aviation Service. Equipment at first consisted of three Blériot aircraft, but soon three more of this type were placed in service, and then three Sanchez-Besa, a Déperdussin and a Voisin aircraft were also added to the strength. In 1915 ten aircraft were being operated in two squadrons. Two years later, six Bristol M.1Cs were bought, and the following year one of these aircraft made the first flight across the Andes. A Chilean aircraft factory and a Naval Aviation Service were both founded in 1919. The Naval Aviation Service operated seaplanes. A few war-surplus D.H.4 bombers were added to the Military Aviation Service's slowly increasing fleet in 1921.

The Fuerza Aérea de Chile was formed in 1930 as a separate service on the merger of the two air arms. Operational equipment of the new air force included Vickers Wibault and Curtiss Hawk III fighters, Junkers R-34 and Dornier bombers, Vickers Vixen and Curtiss Fox general-purpose aircraft, Dornier Wal flying-boats, Fairey IIIF seaplanes, Loening C.2 and Sikorsky S-38 amphibians in a naval co-operation group, and de Havilland Gipsy Moth and Avro 504 trainers. New arrivals included Focke-Wulf Fw 44, Avro 626, and Nardi F.N.305 trainers, and Arado Ar 95 general-purpose aircraft.

The FAC was reorganized in 1941 by an American Aviation Mission, which also provided Curtiss P-40 and Republic F-47 Thunderbolt fighters, Douglas B-24 and North American B-25 Mitchell bombers, Sikorsky OS2U-3 seaplanes and Consolidated PBY-5A Catalina amphibians, Fairchild PT-19, North American T-6 and Vultee BT-13 trainers. Additional Thunderbolts were delivered after the war ended, and during the early 1950s fifty

Beech T-45 Mentor and D18S trainers were bought. Eight de Havilland Canada DHC-2 Beaver light transports were bought in 1952, and in 1957 DHC-3 Otters supplemented these aircraft following a small number of Beech Twin Bonanza transports. The end of the decade saw Chile's first jets, de Havilland Vampire T.55 trainers, arrive, along with some Douglas B-26 Invader bombers and fifty Chilean-built Chincol trainers.

Chile became a founder-member of the Organization of American States in 1948 although she has never been completely dependent on the United States for military supplies.

The 1960s saw a number of second-line aircraft types enter service, including some helicopter types. Twenty Lockheed F-80C Shooting Star fighter-bombers were also obtained, and in 1969 and 1970 these were replaced by Hawker Hunter FGA.9s. A change of government in 1970 is likely to result in very much closer relations with the Communist bloc.

Currently the FAC, which has 8,000 men, operates thirty Hawker Hunter FGA.9 and FGA.71 fighter-bombers; fifteen Douglas B-26 Invader bombers; fourteen Grumman HU-16B Albatross maritime-reconnaissance amphibians; twenty Beech C-45, twenty-five Douglas C-47 Dakota and four DC-6, twenty de Havilland Canada DHC-2 Beaver, twelve DHC-3 Otter and eight DHC-6 Twin Otter transports, five Beech Twin Bonanza and nine 99A light transport and communications aircraft; ten Cessna 180, four O-1 and twenty North American T-6 liaison and AOP aircraft; eleven Bell 47 Sioux, two UH-1D Iroquois, ten Hiller UH-12E and nine Sikorsky UH-19 helicopters; with ten Cessna T-37B, eight Lockheed T-33A, five de Havilland Vampire T.55, forty-five Beech T-34 Mentor and some Hawker Hunter T.7 trainers.

Chilean Navy

The Chilean Navy operates Bell 47G Sioux helicopters and Beech T-34 Mentor aircraft on communications duties. There will also be Westland Wasp helicopters operated on anti-submarine duties from two *Leander*-class frigates on order from the United Kingdom for 1973 delivery.

REPUBLIC OF CHINA (TAIWAN)

Chinese Nationalist Air Force

China today is divided (as are Germany, Korea and Vietnam) into two states, one in the Eastern and one in the Western bloc. Such divisions are not new in China, however, and for most of the present century the country has been split by a number of warring factions. Prior to the establishment of the Chinese Republic in 1911, one or two aircraft were operated, but it was later that a number of regional air arms came into existence. These included that of Yuan Shih-k'ai, formed in 1914 at Shanghai with twelve Caudron G.III and G.IV aircraft and that of Tuan Chi-jui, formed in northern China, which joined the Allies in 1917 although there was no practical benefit. There were also official Chinese Navy and Army air arms.

In 1919 the Chinese Aviation Service was formed with British and American assistance on the amalgamation of the two service air arms. The CAS's first equipment consisted of some sixty Avro 504K trainers delivered in 1920, which were soon joined by forty Handley Page 0/400 bombers and transport conversions of the Vickers Vimy. Morane-Saulnier trainers were also added to the strength, along with Breguet Br.14B-2 and Ansaldo A.300 bombers which were soon operating against insurgents in China's many provinces.

Prior to the Japanese invasion of Manchuria in 1931, the Manchurian Air Arm operated Breguet Br.14B-2, Potez XXV and Handley Page 0/400 bombers, and Caudron C.59 and Schreck F.B.A. flying-boats.

The Chinese Government requested British assistance in forming an air force after Japan invaded Manchuria, but this was refused. An American Aviation Mission provided Boeing 218 and Curtiss Hawk fighters to form a six-squadron Central Government Air Force by 1934. Other aircraft supplied during this period from the United States included Northrop Gamma, Vought V.65 Corsair, and Douglas 0-38 bomber and attack aircraft, and Fleet trainers. In 1935 an Italian Mission brought Fiat C.R.30 and C.R.32, and Breda Ba.27 fighters; Fiat B.R.3, Caproni Ca.111, Savoia-Marchetti S.M.7 and Heinkel He111A bombers; and Breda Ba.25 trainers. There were also a handful of Blackburn Lincock fighters and Vickers Vespa VI reconnaissance-bombers from the United Kingdom. The provincial air arms survived during this period, notable being the Kwantung Air Force and the Kwangai Air Force, operating Russian and British aircraft respectively, although the latter allied itself with the Central Government in 1936.

A further Japanese attack started in 1937 and, in spite of several important battles, the Japanese had very much the advantage in numbers, modern equipment and training. The Soviet Union provided the Central Government Air Force with a number of aircraft, including Polikarpov I-15 and I-16 fighters, Tupolev SB-2 bombers, and a number of aircraft in six squadrons manned by Russian pilots. The Chinese gained air supremacy for a period, until the Japanese moved their latest aircraft to the area and regained their advantage. Japan soon occupied all of central China, Manchuria and some lesser regions.

A number of occupied provinces were permitted to operate their own air arms under Japanese control. The leaders at this time included the Cochin Chinese Air Force in Nanking, operating Nakajima Ki.34 transports and Tachikawa Ki.9 trainers, and the Manchoukuoan Air Force in Manchuria operating Nakajima fighters and Kawasaki bombers, followed by some more modern aircraft, before its personnel were absorbed into the Japanese forces.

The Central Government Air Force had not been inactive. An American Volunteer Corps was formed in 1941 with ninety Curtiss P-40B fighters; known as the Flying Tigers (former members started the airline of the same name after the war,) this force was absorbed into the regular American forces in 1942. Throughout World War II, a considerable number of American aircraft were supplied to the Central Government Air Force, including 129 Vultee 48C, 377 Curtiss P-40B/E, 108 Republic P-43, fifteen Lockheed P-38 Lightning and fifty North American F-51 Mustang fighters; 130 North American B-25 Mitchell bombers; twenty-eight Lockheed A-28 ground-attack aircraft; twenty Curtiss C-46 Commando, eighty Douglas C-47 Dakota and some C-53 transports; twenty North American AT-6, eight Beech AT-7, fifteen Cessna AT-17, 150 Boeing-Stearman

PT-17 Kaydet, 135 Fairchild PT-19, seventy Ryan PT-27 and thirty Vultee BT-13 trainers.

After the war ended, the Central Government Air Force was reorganized and renamed as the Chinese Air Force, with some ex-USAF aircraft to supplement those surviving from the war. The new arrivals included additional Mustangs, Lightnings and Mitchells, with some Republic F-47 Thunderbolt fighter-bombers and Consolidated B-24 Liberator heavy bombers, while 250 ex-RCAF de Havilland Mosquito fighter-bombers were purchased. However, the advance of Communist forces across China forced the Nationalists, who had formed the Central Government, to withdraw to the offshore islands of Formosa and Taiwan in 1949.

The new Chinese Nationalist Air Force had some 160 aircraft, a representative sample in fact of the aircraft which had been entering Central Government and Chinese Air Force service over the preceding five or six years. In 1951, US military aid started, with deliveries of Republic F-84G Thunderjets, while in 1954 the remaining Thunderbolts were replaced by North American F-86F Sabres, and the Lightnings were replaced by Republic RF-84F Thunderflash reconnaissance-fighters. Fifty North American F-100 Super Sabres, followed by Lockheed F-104A Starfighters, were the next types in service. In 1963, Lockheed F-104G Starfighters replaced the Thunderjets and some of the earlier F-104As. The Sabres are in turn being replaced by Northrop F-5A/B Freedom Fighters.

A number of transport aircraft were also provided for the CNAF, including Fairchild C-123 Providers and C-119 Packets, and of course there were the inevitable helicopters and jet trainers.

Currently the CNAF has a personnel strength of 65,000 men, and operates forty-five Lockheed F-104G and eighteen F-104A Starfighter interceptors; twenty-five each of Lockheed RF-104G Starfighter and McDonnell Douglas RF-101 reconnaissance-fighters; ninety North American F-100A Super Sabre and seventy Northrop F-5A fighter-bombers, which have replaced all but a few F-86F Sabres; forty Fairchild C-119 Packet and ten C-123 Provider, thirty Curtiss C-46 Commando and fifty Douglas C-47 Dakota transports; Beech C-46 communications aircraft; Grumman HU-16 Albatross amphibians for rescue duties; six Hughes 500M, a Bell HH-13 and seven Sikorsky

UH-19 helicopters; North American T-6, T-28D and TF-100F, Lockheed TF-104G and T-33A, and Northrop F-5B trainers.

Chinese Nationalist Army

The Chinese Nationalist Army operates Bell UH-1H Iroquois helicopters, introduced in 1970, on light transport and communications duties.

CHINESE PEOPLE'S REPUBLIC

Air Force of the People's Liberation Army

Communist forces took control of mainland China in 1949. It was during the period of 1945-9 that Soviet assistance came to the fore (although during the immediate pre-war period Russia had provided aircraft and pilots) and included the building of a flying school in Manchuria in 1948. At first the Air Force of the People's Liberation Army used former Chinese Air Force aircraft, including North American F-51 Mustang fighters and B-25 Mitchell bombers, and Curtiss C-46 Commando and Douglas C-47 Dakota transports. These were soon supplemented by Yakovlev Yak-9 and Lavochkin La-11 fighter aircraft, and although little part was played by China in the Korean War at its beginning in 1950, Mikoyan MiG-15 jet fighters supplied by Russia in 1951 soon saw action in Korea.

Throughout the 1950s a considerable volume of Soviet aid was provided to China, though relations between the two countries were beginning to chill during the latter part of the decade. The aid included additional MiG-15 fighters, and also MiG-17 and MiG-19 fighters; MiG-21 interceptors; Tupolev Tu-2 and Tu-4 (Boeing B-29 Superfortress copy) and Ilyushin Il-28 (jet) bombers; Lisunov Li-2 (Douglas C-47 copy), Antonov An-2, and Ilyushin Il-12 and Il-14 transports; Mil Mi-1 and Mi-4 helicopters; Yakovlev Yak-12 and Yak-18, MiG-15UTI and Il-28U trainers. China built all of these aircraft in addition to the Soviet supplies, the Mikoyan MiG series becoming the Shenyang F-2 (MiG-15), F-4 (MiG-17), F-6 (MiG-19) and F-8 (MiG-21).

Currently, the AFPLA has some 180,000 men and women operating an unknown number of aircraft, although the Shenyang F-6 predominates, with a few F-2s, F-4s and F-8s in the fighter, fighter-bomber and interceptor roles. The Ilyushin Il-28 is the most common bomber, and there are some Tupolev Tu-4 and Tu-16 aircraft, while on transport duties there are Antonov An-2, Ilyushin Il-14 and Il-18 aircraft, and Mil Mi-4 helicopters. Upwards of 350 aircraft in the Civil Air Bureau's fleet would, however, be available for military duty. China also has some SA-2 Guideline surface-to-air guided missiles but, while there are nuclear weapons available, few up-to-date bombers exist for delivery.

Chinese Navy

The Chinese Navy operates almost 1,000 aircraft, including 350 MiG series fighter-bombers and 80 Ilyushin Il-28 bombers, with helicopters, transports and seaplanes in a supporting role. There are no aircraft carriers and, since there are few ships of more than escort size, few aircraft, if any, can be carried aboard ship. This situation may change in the future, however, since the latest reports indicate that attention is being paid to an increase in naval strength.

COLOMBIA

Colombian Air Force *Fuerza Aérea Colombiana*

Colombia's history of military aviation dates from the purchase of a Caudron G.IIIA biplane in 1922 for a flying school. At about the same time, the Colombian Navy formed a flying-boat flight. Both air arms were fairly modest; amongst the few aircraft operated were Curtiss Hawk fighters and Falcon reconnaissance aircraft; Bellanca 77-140 bombers; Curtiss-Wright CT-32 Condor, Ford 4-AT-E and Junkers W.33 and W.34 transports; with North American NA-16-3, Consolidated PT-11 and Curtiss Fledgling trainers. The Navy operated a handful of Seversky SEV-3MWW amphibians.

In 1943, on the recommendation of an American Aviation Mission, the two air arms were merged to form a separate service, the Fuerza Aérea Colombiana. In 1948, Colombia became a member of the Organization of American States, and started to receive American military aid. This included a squadron of Republic F-47 Thunderbolt fighter-bombers, a squadron of Boeing B-17G Fortress and a squadron of North American B-25 Mitchell bombers, a squadron of Convair PBY-5A Catalina amphibians, and a squadron of Douglas C-47 Dakota transports, with Boeing-Stearman PT-17 Kaydet and North American T-6 Texan trainers. De Havilland Canada DHC-2 Beaver transports arrived in 1951, while in 1954 six Lockheed T-33A jet trainers and several Beech T-34 Mentor trainers replaced the earlier trainers. Operational jet equipment, Canadair CL-13B Sabre Mk. 6 fighters, arrived in 1956.

During the past few years, the FAC has tended to specialize in counter-insurgency (COIN) operations, in common with many South and Central American air forces.

Currently there are 6,000 men, and a combat squadron of North American F-86F Sabres and Lockheed AT-33s plus six Canadair CL-13B Sabre fighters; eight Douglas B-26 Invader bombers; eight Convair PBY-5A Catalina amphibians; Cessna T-41D and T-37C armed-trainers; fifty transport aircraft, including Douglas C-47 and C-54, two Lockheed C-130 Hercules, twelve de Havilland Canada DHC-2 Beaver and some DHC-3 Otter, and Aero Commander 680 aircraft; twenty Bell 47G/D/J Sioux helicopters; two UH-1D Iroquois, six Kaman HH-43B Huskie, twelve Hughes OH-6As and six 269s, and four Hiller UH-23 helicopters; and Beech T-34 Mentor and Lockheed T-33A trainers.

DEMOCRATIC REPUBLIC OF THE CONGO (CONGO-KINSHASA)

Congo Air Force *Force Aérienne Congolaise*

The Force Aérienne Congolaise was formed in 1961 with light aircraft to oppose an attempt by the Katanga province to break away from the Congo after the granting of independence by

Belgium. After the civil war ended, the official and the rebel air arms merged, and a backbone of foreign pilots was thus provided for the Force Aérienne Congolaise. Belgian advisers and instructors were also used at first, but their place was later taken by an Italian Aviation Mission. The force has since been modernized and expanded.

Currently the force, with less than 1,000 men, operates seventeen MB.326GB armed-trainers which were delivered in 1969; between fifteen and twenty North American T-6 Harvard and T-28 Trojan armed-trainers; with Douglas C-47 Dakota and DC-4 transports; de Havilland Canada DHC-4 Caribou transport aircraft; and de Havilland Dove, Beech 18 and various other light aircraft; plus six Sud Alouette III helicopters and some Piaggio 148 trainers. Seven Sud SA.330 Puma helicopters entered service in 1971.

CONGO (BRAZZAVILLE)

Congo Air Force

This former French colony received on independence the standard arms package of a Douglas C-47 Dakota transport, three Max Holste 1521M Broussards and a Sud Alouette II helicopter.

CUBA

Cuban Air Force *Fuerza Aérea Revolucionaria*

Plans for a Cuban military air arm were first made in 1915, but it was not until 1917 that six Curtiss JN-4D trainers were bought. No further aircraft appear to have been purchased until 1923, when six D.H.4B bombers and a number of Vought VO-2 AOP aircraft entered service. A major setback occurred in 1926 when most of the aircraft, operated by what had been named the Aviation Corps, were destroyed in a hurricane. A number of years were to pass before a full recovery could be made, but by 1934, when the Aviation Corps was reorganized into Army Aviation and Naval

Aviation for co-operation duties, Waco D-7 general-purpose biplanes, Bellanca Aircruiser and Howard transports, Stearman A73 and Curtiss-Wright 19R-2 trainers were being operated.

During World War II, Cuba put her bases at the disposal of the Allies, and in return received Grumman G.21 Goose amphibians; Aeronca L.3 AOP aircraft; and Boeing-Stearman PT-13 and PT-17 Kaydet, and North American T-6 trainers. After the war, Cuba became a member of the Organization of American States, and received American military aid including North American F-51D Mustang fighters; B-25J Mitchell bombers; and Douglas C-47 Dakota transports. The first jets, Lockheed T-33A trainers, were delivered in 1955, while in 1957, three de Havilland Canada DHC-2 Beaver transports entered service, followed a year later by two Westland Whirlwind helicopters. A 1955 reorganization resulted in the force being returned wholly to the Cuban Army and renamed the Fuerza Aérea Ejercito de Cuba (Cuban Army Air Force).

A revolution in 1958 resulted in a change of government, with the emphasis away from the West and towards the East – indeed, Russian military advisers and instructors arrived in Cuba – and the present title was adopted. A succession of Russian equipment flowed into Cuba, although the United States successfully prevented inter-continental missile (ICBM) bases being established there.

Currently, Cuba has some 12,000 men in the FAR, which operates twenty Mikoyan MiG-15 fighter-bombers; forty MiG-19 and seventy-five MiG-17 fighters; fifty MiG-21 interceptors; a total of fifty transport aircraft, including Antonov An-2, An-24 and Ilyushin Il-14 types; twenty-five Mil Mi-4 and thirty Mi-1 helicopters; thirty MiG-15UTI, Zlin 226 and other trainers. There are also many SA-2 Guideline surface-to-air missiles.

CZECHOSLOVAKIA

Czechoslovak Air Force   *Ceskoslovenske Letectvo*

Czechoslovakia came into existence in 1918 with the federation of Bohemia, Moravia, Slovakia, Ruthenia, and part of Silesia, and almost immediately a Czechoslovak Army Air Force was formed based on the air units of the Czech Legions which had existed in France and Russia. Ex-World War I aircraft were pressed into service, and the French Government provided assistance, while provision was made for Czech production of French aircraft. Soon the first Czech-designed aircraft, including Aero A-18 fighters, Smolik Sm-1 and Sm-2 bombers and Letov S-10 trainers, were finding their way into CAAF service.

By the end of the first decade of its existence, the CAAF had four hundred aircraft in twenty-five squadrons, and by this time the major types were the Avia B.H.21 and Letov S-20 fighters; Aero A-24 and Letov S-16 bombers; Aero A-11 and A-12 AOP aircraft; Avia B.H.10 and B.H.11, and Letov S-10 and S-18 trainers. During the 1930s the Avia B.H.33 and B.H.34 fighters became standard, along with Potez 63 fighter-bombers, Fokker F.VII and F.IX tri-motor bombers, Aero A-30 and A-100 reconnaissance aircraft, and Aero A-32 AOP aircraft. The Aero B-17 bomber was a licence-built Tupolev SB-2 which entered service just before the start of World War II, along with Avia 135 fighters and Avia A-300 bombers.

In 1938 the Munich Agreement deprived Czechoslovakia of one-third of her territory, and the following year German troops occupied the rest of the country, dismantling the Czechoslovak Republic and making Slovakia an independent state. No resistance was offered to the invader, but Czech personnel who escaped joined the Polish and French Air Forces and were particularly successful while flying with the latter before the fall of France. From 1940 onwards, many Czech personnel flew with the RAF, which operated three Czech fighter squadrons, with Hawker Hurricanes and Supermarine Spitfires, and a bomber squadron, which operated Vickers Wellingtons at first, and later Consolidated Liberators. Other Czechs fought with the Soviet forces. The Germans eventually formed a Slovak Air Force which fought alongside the Luftwaffe, operating Messerschmitt Bf 109G fighters.

After the war ended, the new Czechoslovak Air Force was largely equipped with abandoned Luftwaffe and Slovak Air Force equipment, plus the aircraft flown by the former RAF Czech squadrons, to which a small number of de Havilland

Mosquito fighter-bombers, Lavochkin La-7 fighters, and Petyakov Pe-2 bombers were added.

In 1948 a Communist *coup d'état* was successful. The CAF's personnel were purged of all pro-Western and ex-RAF elements, and Russian advisers moved in, while Soviet forces were stationed in Czechoslovakia. Initial supplies of Soviet equipment included Ilyushin Il-10 ground-attack aircraft, Lisunov Li-2 (C-47 copy) and Ilyushin Il-12 transports, while, in 1951, Mikoyan MiG-15 jet fighters entered Czech service. Eventually almost two hundred MiG-15s entered CAF service. Antonov An-2 light transports and Zlin 226 trainers also appeared in Czech colours at about this time. All of the MiG series of fighters and interceptors have been in Czech service, along with Sukhoi Su-7B ground-attack jets and Ilyushin Il-28 jet bombers.

Currently the CAF has some 18,000 men, and operates 150 Mikoyan MiG-21 interceptors; one hundred MiG-19 fighters; eighty MiG-17 and eighty MiG-15 fighter-bombers; 150 Sukhoi Su-7B ground-attack aircraft; sixty Ilyushin Il-28 bombers; sixty Lisunov Li-2, Antonov An-2 and An-12, Ilyushin Il-14 and Il-14 transports; one hundred Mil Mi-1, Mi-4, Mi-6 and Mi-8 helicopters; 150 L-29 Delfin, and another 150 Zlin 226 and 326, Yak-11, MiG-15UTI and Ilyushin Il-28U trainers.

Czech armed forces are probably not fully effective at present. This follows Soviet intervention in Czechoslovakia's internal affairs in 1968. The move was strongly resented by the Czechs, and, although their country is a member of the Warsaw Pact, they can be counted as one of Russia's less enthusiastic allies at present.

DAHOMEY

Dahomey Air Force *Force Aérienne du Dahomey*

A former member of the French Community still maintaining special links with France, Dahomey operates the standard 'package' of a Douglas C-47 Dakota, three Max Holste 1521M Broussards, and a Sud Aviation Alouette II helicopter, while there is also an Aero Commander 500B for VIP transport duties – the gift of the United States Government.

DENMARK

Royal Danish Air Force *Flyvevaabnet*

The Royal Danish Air Force has the distinction of being Europe's second newest air force, dating from a merger of Danish military and naval aviation in 1950. However, it was in 1912 that the two predecessor services received their first aircraft: an Henri Farman for the Royal Danish Navy, and a Danish-designed B & S Monoplane for the Army. These were followed in 1913 by two Donnet-Leveque flying-boats for the RDN, and an Henri Farman and a Maurice Farman for the Army in 1914. During World War I, Denmark remained neutral and was wholly dependant upon aircraft built at the Royal Army Arsenal. After the war, several ex-German aircraft supplemented the RDN's small force, along with a number of Avro 504 trainers bought in 1920, while the Army was equipped with L.V.G. B.III, Fokker C.I and Potez XV aircraft.

The formation of a Naval Flying Corps and an Army Flying Corps in 1923 marked the next stage of development. In 1926 the Naval Flying Corps was reorganized into two squadrons: No. 1 Luftflotille with Hansa-Brandenburg reconnaissance-seaplanes which were replaced in 1928 by Heinkel He 8 seaplanes; and No. 2 Luftflotille with Hawker Danecocks (licence-built Woodcocks) which were replaced by Hawker Nimrods in 1935. Other aircraft operated by the Naval Flying Corps during the inter-war years included de Havilland Gipsy Moth and Avro 621 trainers, Hawker Dantorp torpedo-bombers, and an ex-Lufthansa (German Airlines) Dornier Wal flying-boat which, with three He 8s was used on survey work in Greenland, ultimately to be of great benefit to the Allies during World War II when establishing a chain of bases for ferrying aircraft across the North Atlantic.

Although both services were further reorganized in 1932, the main effect this time was felt by the Army Flying Corps which became the Havaens Flyvertropper (Army Aviation Corps) with a prospective five squadrons. No. 1 Eskadrille was equipped with Bristol Bulldog fighters; No. 2 Eskadrille with Fokker C.V. reconnaissance aircraft; and No. 3 Eskadrille with Fokker C.I. but in 1934 re-equipped with Fokker C.Vs, which also equipped

No. 5 Eskadrille in 1935. No. 4 Eskadrille was never formed due to defence cuts. Gloster Gladiators replaced No. 1's Bulldogs in 1935.

The overwhelming superiority of the Luftwaffe in 1940 made resistance pointless, and most Danish aircraft were strafed on the ground. Several pilots made their way to the United Kingdom where they joined the Royal Air Force.

After the war the two air arms were re-established, with six Percival Proctor and a couple of liaison aircraft as initial equipment. A committee was established with object of creating an autonomous air force, and the first result of its work was a joint flying school formed in 1946. In 1947 a joint Air Staff was created.

An assortment of aircraft were put into service during the years immediately before the RDAF's formation in 1950: Supermarine Spitfire IX fighters and Sea Otter amphibians; Convair PBY-5A Catalina amphibians; Airspeed Oxfords and North American AT-6 Harvard trainers; and, in 1949, Denmark's first jets, Gloster Meteor fighters.

On its formation in October 1950, the RDAF had five squadrons: Eskadrille No. 721 operating Convair PBY-5A Catalinas, Supermarine Sea Otters and Airspeed Oxfords; Eskadrille No. 722 operating Supermarine Spitfires and Airspeed Oxfords; Eskadrille No. 723 with Gloster Meteor F.4s, replaced in 1952 by Meteor N.F.11 night-fighters; Eskadrille No. 724 with Meteor F.8s; and No. 725 with Spitfire IXs. At this time, twenty-seven de Havilland Canada Chipmunk basic trainers were delivered. A founder-member of the North Atlantic Treaty Organization, Denmark's commitment was initially eight fighter-bomber squadrons for which the United States supplied two hundred Republic F-84E/G Thunderjets and a number of Lockheed T-33A jet trainers. The Spitfires of No. 725 were replaced in 1951, and new squadrons, No. 726–730 were equipped with Thunderjets in 1952 and 1953. Douglas C-47 Dakota transports and Bell 47D Sioux helicopters were also supplied.

During the late 1950s, thirty Hawker Hunter Mk. 51 fighters and ten Republic RF-84F Thunderflash reconnaissance-fighters were introduced, preceding North American F-100D Super Sabres and, during the early 1960s, Lockheed F-104G Starfighters. Other aircraft of the period included ex-RCAF Fairey

Fireflies for target-towing, Hunting Pembroke C.52 communications aircraft, and Sikorsky S.19 helicopters. In 1970, the Catalinas were finally replaced for air-sea rescue duties by Sikorsky S.61A helicopters.

Currently the RDAF has some 11,000 men; two Lockheed F-104G interceptor squadrons; two North American F-100D Super Sabre and one SAAB F-35 Draken fighter-bomber squadrons; one SAAB RF-35 Draken reconnaissance-fighter squadron and one Hawker Hunter ground-attack squadron; there are twenty-three aircraft in each of the two Draken squadrons, sixteen each in the other squadrons. Eight Douglas C-47 and five C-54 transports are operated, while there are eight Sikorsky S-61A helicopters in an air-sea rescue squadron. The RDAF is also responsible for the Army's twelve Hughes 500M helicopters and KZ VII Larks used on AOP and liason duties, and for the Royal Danish Navy's eight Sud Alouette III helicopters. Training equipment consists of Lockheed T-33A jet trainers, de Havilland Canada Chipmunks, and Hunter, Starfighter and Super Sabre conversion trainers. Four each of Nike-Hercules and Hawk surface-to-air missile squadrons are operated.

DOMINICAN REPUBLIC

Dominican Air Corps *Aviación Militar Dominicana*

It was not until the early 1940s that the decision to form an air corps of the Dominican Army was taken. A few Aeronca L-3 AOP aircraft were purchased, plus an assortment of American trainers, including Boeing-Stearman PT-17 Kaydets, Vultee BT-13s, and North American AT-6s. In 1948 an expansion programme was initiated which saw the introduction to Aviación Militar Dominicana service of Bristol Beaufighters, de Havilland Mosquito and Republic F-47D Thunderbolt fighter-bombers and a handful of Boeing B-17G Fortress bombers. These were succeeded within four or five years by many of the present generation of aircraft.

Currently the 3,500-strong AMD operates one squadron of twenty de Havilland Vampire Mk. 1 fighter-bombers; a similar number of North American F-51D Mustang fighter-bombers in another squadron; a squadron of seven Douglas B-26 Invader bombers; two Convair PBY-5A Catalina maritime-reconnaissance amphibians; six Douglas C-47 Dakota, six Curtiss C-46 Commando, three de Havilland Canada DHC-2 Beaver, and three Cessna 170 transports; two Bell 47 Sioux, two Sikorsky H-19, seven Hughes OH-6A, and a Sud Alouette III and two Alouette II helicopters; North American T-6 Harvard and Beech T-11 trainers.

Dominica is a member of the Organization of American States.

ECUADOR

Ecuadorian Air Force *Fuerza Aérea Ecuatoriana*

Ecuador operates a small air force, the history of which dates from 1920 when an Italian Aviation Mission arrived in the country, and the Cuerpo de Aviadores Militares was formed. Initial equipment comprised aircraft of Ansaldo, Aviatik and Savoia manufacture. Little further progress was made until 1935, when the present title was adopted, and a few Curtiss-Wright 16E trainers and an Osprey AOP aircraft were bought. Meridionali reconnaissance aircraft were delivered in 1938. An American Aviation Mission in 1941 provided Fairchild, Ryan PT-20 and North American NA-16 trainers, with the first fighter aircraft, Seversky P-35s. The local German-owned airline was taken over at this time, providing the FAE with Junkers Ju 52/3 transports.

In 1948, Ecuador became a member of the Organization of American States, and American defence aid followed, including Republic F-47D Thunderbolt fighter-bombers, Convair PBY-5A Catalina amphibians, Douglas C-47 Dakota and Beech C-45 transports, and North American T-6 trainers. The first jets were introduced in 1954: twelve Gloster Meteor F.R.9 fighters and six English Electric Canberra B.6 bombers. Bell 47 Sioux helicopters were provided as American military aid in 1960, since when this aid has been extended to include twelve Lockheed F-80C Shooting Star interceptors, two Douglas DC-6B transports, Lockheed T-33A, North American T-28, and Cessna T-41A trainers.

Currently the Fuerza Aérea Ecuatoriana has some 3,500 men, and operates ten Lockheed F-80C Shooting Star interceptors; eight Gloster Meteor F.R.9 fighter-bombers; five BAC Canberra B.6 bombers; two Convair PBY-5A Catalina amphibians; Douglas C-47 and DC-6B, and Beech C-45 transports; three Bell 47 Sioux and a Hiller FH-1100 helicopters; North American T-6 and T-28, Cessna T-41A and Lockheed T-33A trainers; and Cessna 180s on AOP duties for the Army. A Short Skyvan 3M light transport was delivered in 1971.

EGYPT (UNITED ARAB REPUBLIC)

Egyptian Air Force

The Egyptian Air Force is Africa's largest, and also the largest in the Middle East. Old by African standards, although naturally a good deal younger than the South African Air Force, it dates from the formation of the Egyptian Army Air Force in 1932 with five de Havilland Gipsy Moths for training, survey and army co-operation duties. For the first five years of its existence the new air force was organized largely by RAF personnel. Additional equipment took the shape of two batches each of ten Avro 626 biplanes in 1933 and 1934, with a Westland Wessex also in 1934. Later a number of Hawker Audaxes, Avro Ansons and Miles Magisters were supplied. Duties were gradually extended to counter-smuggling activities.

As a result of the Anglo-Egyptian Treaty of 1936, Egypt became a sovereign state, and in 1939 the title of the Royal Egyptian Air Force was assumed when the link with the Egyptian Army was broken. More modern equipment was then purchased, including Bristol Blenheim bombers, some second-hand Gloster Gladiator and Hawker Hart fighters, Westland Lysander and Percival Q-6 army co-operation aircraft, and additional Ansons, to form five squadrons: two fighter and three army co-operation. During the first two years of World War II, the REAF assisted the RAF in protection of the Suez Canal, but in 1941 Egypt was declared neutral and the REAF saw no action at all during the war. However, sufficient Hawker Hurricane and Curtiss Tomahawk fighters were supplied to equip two squadrons and one squadron respectively.

After the cessation of hostilities, the RAF resumed assistance for the REAF, and the post-war years were a period of considerable expansion. New equipment included Supermarine Spitfire VB and IX fighters; Handley Page Halifax 9, Short Stirling and Avro Lancaster heavy bombers (which were only used occasionally due to a shortage of suitable aircrew); North American T-6 Harvard and Miles Magister trainers, Curtiss C-46 Commando, Douglas C-47 Dakota and de Havilland Dove transport aircraft.

In May 1948, Egypt invaded Israel, but fighting ceased the following January and the REAF had little success; as a result, however, an embargo was placed on arms supplies. Even so, orders were placed in 1949 for Macchi C.205 fighters, Fiat G.55B fighter-trainers, de Havilland Vampire FB.5 and Hawker Fury fighter-bombers, and Gloster Meteor F.4 fighters and T.7 trainers, which were delivered. An order for twenty Meteor F.4s to supplement those already in service was banned by the British Government, but in 1953 this was by-passed when the REAF received thirty de Havilland Vampire FB.52s from Italy via Syria. Additional Vampires and Meteors followed the lifting of the ban.

In 1954 the Egyptian monarchy was overthrown, and the 'Royal' prefix dropped from the title.

The following year, Soviet military equipment was ordered from Czechoslovakia, including Mikoyan MiG-15 fighters, Ilyushin Il-28 jet bombers, Ilyushin Il-14 transports, and MiG-15UTI and Yakovlev Yak-11 trainers, but although training was given and deliveries were prompt, few were in operation at the time of the Anglo-French-Israeli attack in November 1956, when most of the aircraft were destroyed on the ground.

Replacement MiG-15s and also MiG-17s were delivered in 1957, with some Ilyushin Il-28s; and EAF personnel were trained either in Egypt by Russian and Indian personnel, or in Czechoslovakia. Additional Soviet equipment delivered during the late 1950s and early 1960s included more MiG-17s, MiG-19s and also Tupolev Tu-16 heavy bombers and Antonov An-12B transports, the latter being shared with United Arab Airlines, and

Mil Mi-4 and Mi-6 helicopters. While still receiving assistance itself, the EAF helped AURI (the Indonesian Air Force) and the Yemen Air Force, which at one time consisted almost entirely of Egyptian personnel, and the Syrian Air Force (mainly during the period when the two countries were united in the UAR). A large air defence system was received from Russia to cover the Aswan Dam and the Suez Canal, along with Cairo and Alexandria, and this included Mikoyan MiG-21 interceptors. The EAF's thirty-one squadrons were virtually all destroyed on the ground in the Arab-Israeli War of June 1967, and the Suez Canal was closed.

Since 1967 the Soviet Union has replaced the aircraft lost by Egypt, and also installed an SA-2 Guideline and SA-3 Goa surface-to-air missile system. Unconfirmed reports indicate that Russian personnel are flying many EAF aircraft.

Currently the EAF has some 20,000 men and operates 150 Mikoyan MiG-21 interceptors; 105 Sukhoi Su-7B and 180 MiG-15 and MiG-17 fighter-bombers; twenty-eight Ilyushin Il-28 and fifteen Tupolev Tu-16 bombers; forty Ilyushin Il-14 and twenty Antonov An-12 transports; forty Mil Mi-4, and numbers of Mi-6 and Mi-8 helicopters; a total of 150 MiG-15UTI, L-29 Delfin and Yak-18 trainers. In addition, some MiG-23 interceptors are reported to be on order, but these are hardly likely to be flown by EAF personnel. Soviet personnel man large numbers of SA-2 Guideline and SA-3 Goa surface-to-air missile units.

EIRE

Irish Army Air Corps

The Irish Army Air Corps was formed in 1922 when Eire was granted independence, to provide cover for ground forces on internal security duties. Equipment operated during the first few years included Avro 504K trainers, Bristol F.2B 'Brisfit' and Martinsyde F.4 Buzzard fighters, and D.H.9 bombers. A number of ex-Royal Air Force bases were taken over by the new air arm. De Havilland Tiger Moths were purchased for training duties, and by 1929 some 160 men were serving in the Irish Army Air Corps.

During the years preceding World War II, Gloster Gladiators, Avro Ansons, 621 Tritons, 626 Prefects and 631 Cadets, a Fairey IIIF, Miles Magisters and a de Havilland DH 84 Dragon were introduced. Although Eire was neutral during World War II, a three-squadron strength was maintained, one each with Gloster Gladiators, Avro Ansons and Supermarine Walrus amphibians, plus a few Vickers Vespa army co-operation aircraft. During the war the Gladiators were replaced by Hawker Hurricane IIs, and the Vespas by Westland Lysanders.

Post-war replacement of aircraft took place in a leisurely fashion. For most of the 1950s the IAAC operated Supermarine Spitfire fighters and trainers, de Havilland Dove communications aircraft, and Vampire jet trainers, with some de Havilland Canada Chipmunk basic trainers. The present fleet of some thirty aircraft includes six de Havilland Vampires, eight BAC Provosts, ten de Havilland Canada Chipmunks, three de Havilland Doves, and three Sud Alouette III helicopters, the last mentioned entering service in 1966.

EL SALVADOR

Salvadorian Air Force *Fuerza Aérea Salvadurena*

Five SAML Aviatik trainers were introduced in 1923 for a Military Aviation Service, and these were followed by small numbers of Waco and Curtiss-Wright Osprey general-purpose aircraft, with some Fleet trainers. In 1939 four Caproni ground-attack aircraft were obtained. After World War II ended, Beech AT-11, Fairchild PT-19, Boeing-Stearman and Vultee trainers were received from the United States and, after El Salvador became a member of the Organization of American States in 1948, the US also supplied Douglas C-47 transports and North American T-6 trainers.

Currently the Fuerza Aérea Salvadurena has some 1,000 men, and operates one squadron of six Vought F4U Corsair and another squadron with six North American F-51D Mustang fighter-bombers; four Douglas C-47 Dakota transports; North American T-6, Beech AT-11 Kansan and T-34 Mentor trainers.

ETHIOPIA

Imperial Ethiopian Air Force

Ethiopian military aviation started in 1930 with six Potez 25 bombers and three de Havilland Gipsy Moth trainers under the name of Imperial Ethiopian Aviation. This force was unable to counter the Italian invasion of 1935, which had the support of a relatively strong and well-equipped Regia Aeronautica. Although Ethiopia was liberated by British forces in 1941, it was not until 1946 that the Imperial Ethiopian Air Force was formed with help from a Swedish aristocrat.

The IEAF's first aircraft were ten de Havilland Tiger Moth trainers, but these were soon replaced by thirty SAAB 91A-Safirs, and a few Cessna AT-17 Bobcat trainers. Thirty SAAB-17A bombers were delivered in 1948, with a repeat order during the early 1950s, and eight Fairey Firefly fighters. Stinson L-5 AOP aircraft and Douglas C-47 transports also arrived at this time. The United States started to provide Ethiopia with military aid in 1960, initially with twelve North American F-86F Sabre fighters and some Lockheed T-33A armed-trainers. Re-equipment during the late 1960s included the latest version of the SAAB Safir (the 91C and 91D), Northrop F-5A fighter-bombers, and de Havilland Dove and Ilyushin Il-14 transports.

Currently the IEAF has some 3,000 men, and operates one Northrop F-5A fighter squadron with eight aircraft; twelve North American F-86F Sabres in one fighter-bomber squadron; one BAC Canberra B.2 bomber squadron with six aircraft; two ground-attack squadrons, one with eight SAAB-17A, and one with six North American T-28 and three Lockheed T-33A; an Ilyushin Il-14, six Douglas C-47 and two C-54, and three de Havilland Dove transports in one squadron; three Sud Alouette III helicopters and an Agusta-Bell 204B; and SAAB 91C/D Safir, North American T-28 and Lockheed T-33A trainers.

FINLAND

Finnish Air Force   *Ilmavoimat*

Finland is a small neutral country bordering the Soviet Union, and until 1917 when the Finnish Parliament took advantage of the upheavals of the Russian Revolution to declare independence, the country was controlled by Russia. The war which followed saw the birth of Finnish military aviation with initially a Swedish-built Albatros reconnaissance-bomber, and then a Friedrichshafen seaplane, a flying-boat, additional Albatros aircraft, Nieuport 12s and 17C-1s, and Morane Parasols. Many of these aircraft were flown by Swedish volunteers.

A Treaty in 1920 marked the end of the war. The Finnish Flying Corps, which had been the name given to the air arm, became a separate service, the Ilmavoimat, or Finnish Air Force. French training assistance was received, and the Finnish Air Force received twenty Breguet Br.14B-2 reconnaissance-bombers; twelve Georges-Levy flying-boats; Caudron G.III, C.59 and C.60 trainers; plus more than one hundred licence-built Hansa-Brandenburg A.22 seaplanes. The mid-1920s saw a small number of Gourdou-Leseurre C.1 and Martinsyde F.4 Buzzard fighters enter service, and a British Aviation Mission was sent to Finland in 1924.

The inter-war years saw no improvement in relations between Finland and Russia, which wanted to use Finnish bases. In the meantime a succession of aircraft entered service, including Blackburn Ripon II reconnaissance and torpedo seaplanes; Gloster Gamecock II and Bristol Bulldog fighters; de Havilland and Letov trainers; and licence-built Fokker D.XXI fighters and C.X reconnaissance aircraft; with the nationally-designed Viima and Tuisku trainers. Germany detained thirty-five Fiat G.50 fighters while on their way to Finland, but before the war with Russia broke out in 1939, eighteen Bristol Blenheim bombers had been delivered.

At first, Finland held its own against some 900 obsolescent Soviet aircraft, but this force had grown to some 2,000 by 1940 when Gloster Gladiator, Hawker Hurricane, Brewster 239, and Curtiss Hawk 75A and A-4 fighters arrived, plus some Westland

Lysander army co-operation aircraft and additional Bristol Blenheim bombers. However, hopelessly outnumbered, Finland was forced to surrender in late 1940. The following year, Finland allied herself with Germany against Russia, and in return received Morane-Saulnier M.S.206 and Messerschmitt Bf 109G fighters, Junkers Ju 88A and Dornier Do 17 bombers, and Dornier Do 22W seaplanes. Finland was again forced to surrender in 1944, and afterwards was at war with Germany.

After the war the Finnish Air Force received sufficient Messerschmitt Bf 109G fighters to standardize on the type with four squadrons. The Treaty of Paris in 1947 restricted the Finnish Air Force to sixty aircraft and 3,000 men. In 1955, Finland received its first jets, de Havilland Vampire F.B.52s, along with two Hunting Pembroke C.53s for communications and photographic duties. In 1958, Folland Gnat fighters were introduced, and shortly afterwards licence-production of Potez Magister jet trainers started.

Currently the Finnish Air Force has some 3,000 men and operates twenty Mikoyan MiG-21 interceptors in two squadrons, and one squadron of nine Folland Gnat fighters; sixteen Potez Magister armed-trainers; ten Douglas C-47 Dakota, and two Hunting Pembroke C.53, and a few de Havilland Canada DHC-2 Beaver transports; four Mil Mi-4 and two Mil Mi-8, two SM-1 (Polish-built Mil Mi-1), a Bell JetRanger and two Sud Alouette II helicopters; thirty SAAB-91C Safir, fifty-five Potez Magister, and a number of MiG-15UTI and MiG-21UTI trainers.

FRANCE

French Air Force   *L'Armée de l'Air*

France has a long history of aviation activity, and so while not a late arrival in military aviation, it is perhaps surprising that the origins of the French Air Force only date from 1910. In that year the French Army received a Blériot, two Farman and two Wright aircraft. The name of the new air arm was the Aviation Militaire, and by the following year no less than thirty aircraft and balloons were being operated, although some were being flown by naval officers pending the formation of a naval air arm, which was called the Service Aéronautique. The squadron, or or *escadrille*, was introduced in 1912. This period also saw experiments in aerial photography and radio transmission.

The outbreak of World War I in 1914 found the Aviation Militaire with twenty-one squadrons in France, five with Maurice Farmans, four each with Henri Farmans and Blériots, two each with Voisins and Déperdussins, and one each with Breguets, Caudrons, Nieuports and R.E.P.s. Another four squadrons were in the colonies. During the early stages of the war, the main activity was reconnaissance, although it was by no means unknown for an observer to drop a 90mm. shell fitted with fins over the side of the aircraft! Towards the end of 1914 the Voisin squadron became solely employed in bombing duties but, although there had been battles in that year, it was not until 1915 that aerial combat really started. Morane-Saulnier Type Cs formed the first fighter squadrons of the Aviation Militaire.

Equipment of the Aviation Militaire during the closing stages of the war included Caudron G.III and Maurice Farman AOP aircraft; Caudron G.IV, Dorand A.R.-1s and Voisins on reconnaissance duties; and Breguet-Michelin IV, licence-built Caproni tri-motors, and Breguet Br.14 bombers; with Morane-Saulnier, Spad S.7C and S.13C, Nieuport 17C-1 Bébé and Caudron R.XI fighters. At the end of the war the total strength included some 1,600 AOP and reconnaissance aircraft in 140 squadrons, 480 bombers in thirty-two squadrons, and 1,400 fighters in eighty-three squadrons. However, some sixty per cent of the total strength were casualties, the highest of any Allied service.

Naturally the return of peace brought a reduction in strength, to 180 squadrons based in France, Germany, Algeria, Tunisia and elsewhere in the French African colonies. In 1920 a squadron was formed in French Indo-China operating Breguet Br.14 bombers. New aircraft, Breguet Br.16 bombers and Nieuport 29C fighters soon started to supplement and replace wartime equipment. For the most part, the Aviation Militaire's inter-war duties consisted of police and counter-insurgency duties in the colonies, but in 1925 a rather more serious uprising was led by Abd el Krim in Morocco, which even with the reinforcements sent from France lasted until 1934.

During the 1920s, new aircraft entering service included the Lioré-Gourdou-Leseurre 32, Nieuport-Delage 62C, Spad 81, Wibault, Lioré-et-Olivier LeO 20, Amiot 122, and Blériot 127M fighters; Potez 25 reconnaissance aircraft; Morane-Saulnier M.S.35, 130 and 138, Hanriot-Dupont 32 and Caudron C.59 trainers. These were followed during the 1930s by Nieuport-Delage 629C and Morane-Saulnier M.S.225C-1 fighters; Lioré-et-Olivier LeO 206 four-engined bomber and Potez 39 AOP aircraft. Most significantly, however, the Aviation Militaire, which had been hitherto the 'fifth arm' of the French Army (the others were the infantry, cavalry, artillery and engineers), became the Armée de l'Air, and gained control of six shore-based naval squadrons, including two with Dewoitine D.500 fighters.

Defence expenditure does, however, always receive a low priority in peacetime, and because of this and the unsettled French political situation, there was some neglect of the Armée de l'Air. However, in 1935 an urgent modernization and expansion programme was started, with deliveries of Dewoitine D.500, D.501 and D.520 fighters; Morane-Saulnier M.S.406 and Bloch M.B.151 fighters; Farman F.221, Bloch M.B.210 and Lioré-et-Olivier LeO 45 bombers; plus some Bloch and Potez multipurpose aircraft. Even so, expansion did not keep pace with the deteriorating international situation since nationalization of the French aircraft industry in 1936 and 1937 left it in a chaotic state. Foreign aircraft were ordered, mainly from the United States, but there was insufficient time for these orders to be fulfilled.

At the outbreak of World War II in 1939, obsolete aircraft were transferred to areas of low risk, although a few had to be used for leaflet dropping. There was some success in conflict with the Luftwaffe, but German successes on the ground and their overwhelming aerial superiority meant that France was forced to surrender in 1940, in spite of British assistance in the form of an expeditionary force and RAF units. Germany intended to disband the Armée de l'Air immediately, but after the British attack on the French fleet at Mers el Kébir to prevent it falling into Axis hands, she granted a stay of execution in the hope that the Vichy French would allow it to support the Axis powers. The force was finally disbanded in 1942 after some units in North Africa defected to the Allies.

The Free French Air Force was formed within months of France's fall, initially with Supermarine Spitfire and Hawker Hurricane fighters, Bristol Blenheim and Martin Maryland bombers, and Westland Lysander army co-operation aircraft, plus a few 'refugee' French aircraft. Operations started in 1942, and the following year a merger with former Vichy units led to the Armée de l'Air being re-formed. Other wartime equipment included North American F-51 Mustang, Republic F-47 Thunderbolt, Bell Airacobra, Lockheed Lightning and de Havilland Mosquito fighter-bombers; Douglas Dauntless, Lockheed Lodestar and Handley Page Halifax bombers; and the inevitable Douglas C-47 Dakota transport.

Post-war France had immediately to quell a number of insurrections. In French Indo-China, personnel arrived before their Spitfires, so they operated ex-Japanese Nakajima Ki 43 fighters.

During the late 1940s a process of modernization and standardization began with orders for de Havilland Vampire F.1 jet fighters, which were produced under licence in France, while production of the Dassault M.D. 450 Ouragan was also put in hand. A founder-member of the North Atlantic Treaty Organization, France qualified for American military aid, including Republic F-84F/G Thunderjet fighter-bombers and Lockheed T-33A jet trainers. Aircraft in service during the mid-1950s included the Thunderjets, Grumman Bearcats, Gloster Meteor N.F.11 night-fighters; Morane-Saulnier M.S.500s, Nord 1100 (a development of the Messerschmitt Bf 108), Beech C-45s, Bell 47 Sioux and Sikorsky S-55 helicopters. Strength was 123,000 men and some seventy-five squadrons. New aircraft of the period included the Sud Vautour fighter-bomber, Dassault Mystère fighter, the Nord Noratlas transport, and the Morane-Saulnier M.S.733 Alcyon trainer. Mystères and Thunderjets flew several operational sorties at the time of the Suez crisis of 1956.

Morane-Saulnier M.S.760 Paris and Potez Magister jet trainers replaced the Lockheed T-33As during the latter half of the 1950s and during the early 1960s the Dassault Etendard IV fighter-bomber and the highly successful Dassault Mirage IIIC interceptor first appeared in service. The mid-1960s saw the introduction of the Mirage IVA nuclear bomber to carry French atomic weapons which became available at that time; the Franco-German Transall transport followed towards the end of the decade, replacing the Noratlas. Sud Alouette II and III helicopters first appeared during the mid-1950s and entered service shortly afterwards. Most French helicopter production has since been directed towards the needs of the Army and Navy.

Current personnel strength of the Armée de l'Air is 105,000 men, and organization is based on four commands: Strategic Air Command, Air Defence Command, Tactical Air Force (really two commands grouped together), and Air Transport Command. Air Defence Command includes three squadrons of Dassault Mirage IIIC interceptors, two squadrons with Mirage IVA fighters, three squadrons with Dassault Super Mystère interceptors and two squadrons with Sud Vatour fighter-bombers, which are being replaced by 105 Dassault Mirage F.1 interceptors on order for 1971 and 1972. Strategic Air Command has nine squadrons with a total of forty-five Dassault Mirage IVA nuclear bombers, and three squadrons with a total of twelve Boeing KC-135F tanker aircraft, plus responsibility for two squadrons of surface-to-surface missiles. The Tactical Air Force operates three North American F-100D Super Sabre squadrons, nine squadrons with Dassault Mirage IIIE and two with Mirage IVA fighter-bombers, plus three Mirage IIIR and IIIRD reconnaissance squadrons. Some two hundred BAC-Breguet Jaguar strike aircraft will enter Tactical Air Force service during the early 1970s. Two Douglas A-1D Skyraider squadrons are based overseas. Air Transport Command operates some fifty-six Transall transport aircraft, some Douglas C-47s and DC-6s, Breguet Br.765 Saharas, and several Sikorsky H-34 and Sud Alouette II and III helicopter squadrons. Six BAC Canberra B.6 jet bombers are employed on special duties. Training is on Potez Magister and Morane-Saulnier M.S.760 Paris aircraft, with some M.S.733 Alcyons, and Dassault M.D.315 Flamants. During the mid-1970s 130 Dornier-Dassault-Breguet Alpha jet trainers will enter service.

French Naval Aviation *Aéronautique Navale – L'Aéronavale*

French naval aviation dates from 1910, when the Service Aéronautique was formed with an Henri Farman and a Voisin seaplane. Additional aircraft of Breguet and Nieuport manufacture were put into service before World War I started, and the cruiser *La Foudre* was converted to act as a seaplane tender after manoeuvres in the summer of 1913 helped to prove the value of aircraft for reconnaissance duties.

During World War I, the ex-German cargo vessels *Ann Rickmers* and *Rabenfell* were used as seaplane tenders. The French Navy's aircraft reconnoitred the Austrian Fleet in the Adriatic, and also flew reconnaissance missions against the Turks in Egypt for the British Army. Maximum wartime strength amounted to 1,260 aircraft. Before the end of the war a platform was built over the forward gun turrets of the battleship *Paris* for take-off trials.

The Service Aéronautique was renamed the Aéronautique Maritime in 1925, and the French coastline was divided into six Districts Maritimes, being reduced to four in 1929, amongst which shore-based aircraft were divided. Perhaps a more notable event in 1925 was the *Béarn*, France's first aircraft carrier, joining the fleet. During the late 1920s and early 1930s, support was provided for ground forces fighting in Morocco. Equipment operated during this period included Farman Goliaths, C.A.M.S. 37s and 55s, Latham 43s, Gourdou-Leseurre GL-32s and Dewoitine D.1s, Levasseur P.L.7B and P.L.10Rs operated from the *Béarn*; while the *Commandant Teste* seaplane tender, completed in 1929, operated Latécoère 29s and Gourdou-Leseurre GL-810s. In 1933 six land-based squadrons, including two with Dewoitine fighters, passed to the Armée de l'Air.

Dewoitine D.373 fighters replaced Wibault 74s which had succeeded the D.1s on the *Béarn*, but otherwise the carrier's aircraft remained unchanged until after the outbreak of war, when they were land-based and the *Béarn* was used to ferry aircraft from the United States. New seaplanes and flying-boats were, however, put into service during the 1930s, and these included Lioré-et-Olivier LeO 257, Levasseur P.L.15, Latécoère 298, and Lioré 210 seaplanes; licence-built Short Calcutta, known as the Breguet Br.521 Bizerte, and Lioré 70 flying-boats. Two aircraft carriers, the *Jeffre* and the *Painlevé*, were begun but never completed.

Initially, some successes were scored by the shore-based aircraft against Germany and Italy during that period of World War II which preceded the fall of France. A number of aircraft and pilots escaped to Britain when France surrendered, but units in North Africa attacked units of the Royal Navy and Gibraltar after the Royal Navy had sunk the French fleet at Mers el Kébir. The Vichy-administered French forces in North Africa were divided after the Allied invasion of the area. Aéronautique Maritime squadrons which left Vichy for the Allies were attached to the Armée de l'Air.

In 1945 the Royal Navy loaned the French Navy a light fleet aircraft carrier, which was renamed *Dixmude*, and later the same year the French also obtained HMS *Colossus*, which was renamed *Arromanches*. These ships were invaluable during the war which took place in French Indo-China because shore bases rapidly became unusable. Aircraft operated during this period, and on into the early 1950s included Douglas Dauntless and Curtiss Helldiver dive-bombers; Convair P4Y-2 Privateer and Bloch M.B.175 bombers; Lockheed P2V-6/7 Neptune and Avro Lancaster maritime-reconnaissance bombers; Short Sunderland and Dornier Do 24 maritime-reconnaissance flying-boats; Grumman JRF-5 Goose and Consolidated PBY-5A Catalina amphibians; and Grumman F6F Hellcat, Chance-Vought F4V and AV-1 Corsair fighter-bombers; and Grumman TBM-3 Avenger anti-submarine aircraft. The first jets were Sud Aquilans, which were licence-built de Havilland Sea Venom all-weather jet fighters. North American SNJ-5 trainers were used, and there were Vertol H-22 and HUP-2, Bell 47 Sioux and Sikorsky S-55 helicopters. Two more aircraft carriers entered service, the *La Fayette* and the *Bois-Belleau*. The late 1950s saw seventy-five Breguet Br.1050 Alizé anti-submarine aircraft enter service, mainly operating from the carriers, and one hundred Dassault Etendard strike aircraft.

Today. L'Aéronavale is a small but highly efficient force centred on the aircraft carriers *Clémenceau* and *Foch*, with the *Arromanches* as a combined aircraft and helicopter carrier, and the *Jeanne d'Arc*, ex-*La Resolute*, as another helicopter carrier. There are thirty squadrons of aircraft. Two squadrons operate LTV F-8E Crusader carrier-borne interceptors; BAC-Breguet Jaguar strike aircraft, of which there will be seventy, are replacing Dassault Etendard strike and reconnaissance fighters; there are sixty Breguet Br.1050 Alizé anti-submarine aircraft, forty Breguet Br.1150 Atlantic and some surviving Lockheed Neptune shore-based maritime-reconnaissance aircraft, a dozen or so Sud Super Frelon anti-submarine heavy helicopters, and numbers of licence-built Sud-Sikorsky H-34 helicopters, and Sud Alouette II and III helicopters.

French Army Aviation *Aviation Légère de l'Armée de Terre*

French Army Aviation was formed in 1954 to operate light aircraft and helicopters in support of ground forces. The force proved its worth during the Algerian emergency. ALAT has more than 1,000 aircraft (of which at least six hundred are helicopters), with its organization based on a quota of about forty aircraft per Army division. Equipment includes Sud Alouette II and III, SA.330 Puma and SA.341 Gazelle, Vertol H.21, Sud Super Frelon, and some Bell 47 Sioux helicopters. There are also Max Holste 1521M Broussard (Bushranger), Piper and Cessna O-1 light aircraft, with Nord 3200 and 3400 trainers. It is likely that some Westland WG.13 Lynx helicopters will also enter service within a fairly short time.

GABON

Gabon Air Force *Force Aérienne Gabonnaise*

Originally part of French Equatorial Africa, Gabon became independent in 1960 as a member of the French Community. Only two aircraft are operated, and both are Max Holste 1521M Broussards.

FEDERAL GERMAN REPUBLIC

West German Air Force *Luftwaffe*

The German Military Board purchased its first Zeppelin dirigible in 1907, and in 1910, having purchased further dirigibles in the meantime, obtained the Army's first aircraft, five Henri Farmans, five Wrights, and an Antoinette; while the Naval Air Service was formed with two Curtiss seaplanes the following year. In 1912 the Army's air arm became the Military Aviation Service, and during the years preceding World War I both air arms were expanded. The NAS bought Rumpler-Etrich Taubes, Euler biplanes and licence-built Farmans, to a total of thirty-six aircraft by 1914. The MAS also favoured the Taube, versions of which were built by Albatros, Aviatik, AEG, DFW, Euler, Gotha, LVG, Otto and the Jeannin concern, which built the all-steel Stahltaube, so that half the 1914 strength of 250 aircraft were Taubes of one sort or another.

As elsewhere, aircraft were used primarily for reconnaissance duties at the start of the war, although small bombs were dropped on Paris in August 1914. The Zeppelins proved vulnerable on reconnaissance duties and were diverted to bombing duties, including targets in London and the UK east coast ports, which caused Britain to deploy aircraft for home defence. Germany had thus gained an aerial advantage, and increased this with the introduction of the Fokker E.I. which fired a synchronized machine-gun through the propeller disc. However, at the end of 1915 the Allies restored the balance of air power with the introduction of the Nieuport 17C-1 Bébé and the D.H.2. A further step which gave a period of German supremacy was the introduction of large formations of fighter aircraft, known as a 'circus', in 1916, the most famous of which was led by Baron von Richthofen. Throughout the war the NAS was active with raids on the Kent Coast of England, and the commerce raider *Wolf* used a seaplane for reconnaissance.

Apart from the Taubes, the MAS operated Fokker E.I. E.II, E.III and D.III, Albatros D.I. D.II, and D.III, and Pfalz D.III and Roland D.II fighters; Albatros C.III, Aviatik C.II and LVG C.III reconnaissance-bombers; and AEG G.IV, Friedrichshafen G.III and Gotha G.IV twin-engined bombers. The NAS operated seaplanes of Gotha, Sublatnig, Brandenburg and Friedrichshafen manufacture. Production of aircraft rose from 1,350 in 1914 to a peak of 19,750 a year in 1917. Aircraft and personnel were provided for the air arms of other members of the Central Powers. New types entering service in 1917 and 1918 failed to shake an overwhelming Allied air supremacy at the closing stages of the war.

In 1918 the MAS had some 4,000 aircraft, plus another 15,000 under construction, and 80,000 personnel. Between September 1915 and September 1918, there were 11,000 casualties and more than 2,000 aircraft were destroyed.

The Treaty of Versailles, signed in June 1919, ended German military aviation, and the following year both the MAS and the NAS were disbanded and their aircraft sold or scrapped. Germany was forbidden to construct military aircraft, and until 1926 was also restricted as to the size of civil aircraft which could be built.

In March 1935 the Luftwaffe was founded as an autonomous air force, using a nucleus of former World War I pilots secretly retained after the war, and a number of pilots who had been trained at a military aviation centre established in Russia in 1928. The German aircraft industry had kept abreast of developments by establishing factories in Russia, Sweden and Turkey, while the numerous flying clubs which had thrived in Germany between the wars had given flying experience to many. Initially the new air force was equipped with Heinkel He 51 fighters; and He 45 and 46 reconnaissance aircraft; Junkers Ju 52/3M transports; Dornier Do 11 and Do 23 bombers; and Focke-Wulf Fw 44 Stieglitz and Arado Ar 66 trainers. Some 2,000 aircraft were soon provided, and manufacturers were given loans to expand production and to attract non-aviation engineering firms to build aircraft, of which Blohm und Voss was an example.

The Luftwaffe was divided into area groups. A research centre was formed to evaluate new aircraft, many of which were to play an important part in World War II, including Messerschmitt Bf 109 and Bf 110 fighters, Junkers Ju 87 Stuka dive-bombers, and Dornier Do 17 and Heinkel He III bombers. A splendid opportunity for the Luftwaffe to test its new equipment in combat occurred in 1936 when the Spanish Civil War began, and the forces of General Franco were provided with some twenty Junkers Ju 52/3Mg convertible bomber-transports and six Heinkel He 51 fighters, which were soon supplemented by Heinkel He 70 reconnaissance aircraft and He 59 and He 60 seaplanes which were flown by the Legion Condor, a force formed in Germany. In 1937, Messerschmitt Bf 109B/C fighters replaced the He 51s to counter the new Russian fighters which were appearing in Spain, while, in 1938, Heinkel He 111B and Dornier Do 17E bombers replaced the Ju 52/3Mgs. The Legion Condor returned to Germany in 1939 after the Spanish Civil War had ended, and after numbers of Luftwaffe personnel had received realistic combat training.

In the occupation of Austria in 1938 and of Czechoslovakia in 1939, the Luftwaffe played an important part by providing sufficient aircraft to make resistance appear pointless. The Austrian Air Force was merged with the Luftwaffe, and after a 1939 reorganization replaced the air groups (or *Gruppenkommandos*) with air fleets *(Luftflotten)*, one of these was commanded by an Austrian officer. Each *Luftflotte* had two or three – *Luftgou* (air districts). There were also *Gruppe* and *Staffelnen* – groups and squadrons of about the same individual strength as their RAF counterparts. At the outbreak of World War II in September 1939, the Luftwaffe had a front-line strength of almost 4,000 aircraft, including some 1,300 Messerschmitt Bf 109 series fighters, some 350 Junkers Ju 87 Stuka dive-bombers and 1,300 Dornier Do 17 and Heinkel He 111 bombers.

On 1 September 1939, Germany invaded Poland, and the Luftwaffe gained air supremacy at some cost. The following year saw the spring campaigns against Denmark and Norway, followed by the Low Countries and France, which was occupied by the end of June. Invasion of the British Isles, due to take place in August 1940, was postponed indefinitely because the Luftwaffe, although numerically superior, met its match in the Royal Air Force. Although not quite as fast as the Bf 109, the RAF's Supermarine Spitfires were far more manoeuvrable. In the blitzkrieg which followed, from September 1940 to May 1941, the Luftwaffe continued to suffer heavy losses, notably on the early daylight raids. Malta was next on the list, and was attacked from late 1940 onwards, followed by Yugoslavia, Greece, Crete and Russia. Other Luftwaffe units were already fighting in North Africa.

New aircraft entered service, including Messerschmitt Bf 110 and Focke-Wulf Fw 190A fighters; Dornier Do 217 bombers; Blohm und Voss reconnaissance flying-boats; Focke-Wulf Fw 200 and Condor reconnaissance-bombers, and Heinkel He 115 seaplanes. Towards the end of the war came new weapons, including the FX radio-controlled missile and the Hs 293 glider-bomb, both launched from Do 217 bombers, but more notably the Messerschmitt Me 262 jet fighter introduced in 1944, the Arado Ar 234 jet reconnaissance aircraft, the Messerschmitt Me 163B Komet rocket-powered fighter, and a few of the twin-engined Heinkel He 219A fighter; all of which caused the Allies some concern, but really arrived too late, and in too small a number due to bombing raids on the factories, to affect the outcome of the war. There were also, of course, the V.1 and V.2 missiles used on attacks on London.

April 1945 saw the Luftwaffe grounded through a lack of fuel, its personnel being used as ground troops during the last few months of the war. The Luftwaffe was disbanded immediately the war ended.

Germany was divided into four Allied Zones of Occupation after the war, with the former capital, Berlin, treated similarly. However, the three Western zones were amalgamated to form the Federal German Republic, and in 1955 the Luftwaffe was resurrected. The first aircraft, thirty Piper Cub trainers, were delivered in 1956, followed soon afterwards by a similar number of North American T-6 Harvard trainers and some Lockheed T-33A jet trainers.

West Germany became a member of the North Atlantic Treaty Organization in 1955 and qualified for deliveries of combat aircraft under the American Military Aid Programme. These included 450 Republic F-84F Thunderstreak fighter-bombers; one hundred RF-84F Thunderflash reconnaissance-fighters; Canadair CL-13B Sabre 6 and Fiat F-86K Sabre fighters; Hunting Pembroke, de Havilland Heron, twenty-eight Douglas C-47 Dakota, and licence-built Nord Noratlas transports; Potez Magister jet trainers and Piaggio P.149D trainers; Bell 47G Sioux, Vertol H-21H and H-25, Sikorsky H-34, Saro Skeeter and Bristol Sycamore, and Hiller UH-12C helicopters; and Dornier Do 27B liaison aircraft.

During the early 1960s, licence-built Lockheed F-104G Starfighters and Fiat G.91s replaced the Thunderstreaks, Thunderflashes and Sabres.

Currently the Luftwaffe, with 104,000 men, operates four

Lockheed F-104G Starfighter interceptor squadrons; four reconnaissance squadrons with eighty-eight McDonnell Douglas RF-4E Phantom IIs which replaced RF-104G Starfighters; another four reconnaissance squadrons with Fiat G.91Rs, ten Lockheed F-104G Starfighter fighter-bomber squadrons; and eight squadrons of Fiat G.91R.3 fighter-bombers; six C.160 Transall (Franco-German manufacture) transport squadrons, with a total of about one hundred aircraft, which replaced the Noratlas transports; four Boeing 707-320C jet transports; a variety of Douglas transports; Lockheed Jetstar VIP transports; 130 Dornier-Bell UH-10 helicopters, a few Dornier Do 28 utility transports; Piaggio P.149. Lockheed T-33A, Cessna T-37 and Northrop T-38 Talon trainers. A further 210 McDonnell Douglas F-4 Phantom II are planned for delivery by 1973, while during the mid-1970s the Luftwaffe will receive two hundred Panavia 200 Panther multi-role combat aircraft and two hundred Dornier-Dassault-Breguet Alpha Jet trainers.

German Naval Air Arm *Marineflieger*

The Marineflieger was formed in 1957 for shore-based maritime air operations in the Baltic and the North Sea. A Naval Air Arm had existed during the 1930s operating shore-based Heinkel He 59s and He 60s, Dornier Do 18 flying-boats and Heinkel 115 seaplanes on reconnaissance and minelaying duties, but, after World War II began, this force was merged with the Luftwaffe which had demanded control of all German military aviation. An aircraft carrier, the *Graf Zeppelin,* was under construction as the war ended.

Initial equipment for the Marineflieger in 1957 consisted of Hawker Sea Hawk strike aircraft, Fairey Gannet anti-submarine aircraft, Hunting Pembroke transports and Bristol Sycamore helicopters. These were replaced by Lockheed F-104G Starfighters, Breguet Atlantics and Sikorsky CH-34 helicopters. Currently there are four squadrons each with eighteen Lockheed F-104G Starfighter fighter-bombers; two squadrons each with twenty Breguet Br.1150 Atlantic maritime-reconnaissance aircraft; one squadron operates twenty-three Sikorsky CH-34 helicopters on search and rescue duties which are currently being replaced by twenty-two Westland S-61N Sea King helicopters; while there are also eight Grumman HU-16 amphibians; twenty Dornier Do 28 Skyservants and some Do 27s; a few remaining Hunting Pembroke transports; and Potez Magister jet trainers.

Army Air Corps *Heeresflieger*

The Heeresflieger was formed in 1957 to provide liaison, communications and AOP facilities for the Federal German Army. Altogether some two hundred Dornier Do 27 light aircraft are operated with more than four hundred helicopters, including forty-five Agusta-Bell 47 Sioux and two hundred Bell UH-1D Iroquois, eighty-two Sikorsky CH-53A (S-65), and 225 Sud Alouette II. Eighteen North American OV-10Zs are used for target-towing duties. Air Corps units are attached to Army Divisional HQs.

GERMAN DEMOCRATIC REPUBLIC

East German Air Force
Luftstreitkräfte und Lufverteidigung (LSK)

The German Democratic Republic came into existence in 1949 – the same year as the Federal Republic – and is in fact the former Russian zone of Germany. In 1950 an air arm of the People's Police (Volkspolizei) was formed with Yakovlev Yak-18, Polikarpov Po-2 and Fieseler Fi 156C Storch light aircraft, and former Luftwaffe personnel. In 1955 the Luftstreitkräfte was formed, based on Volkspolizei personnel and aircraft, and with numerous 'flying clubs' and their 'members' as a basis for expansion. Russia supplied Mikoyan MiG-15 fighters and MiG-15UTI trainers, and by 1960 these aircraft and MiG17s formed the LSK's combat strength, supplemented by Ilyushin Il-14M and Tupolev Tu-104 transports. Training was provided in the USSR and in Czechoslovakia, as well as in East Germany. East Germany was, of course, and still is, a Warsaw Pact member.

The 1960s saw a succession of MiG series aircraft enter LSK service, while Mil helicopters were introduced.

Currently the LSK has 21,000 men, and operates eleven MiG-21 interceptor squadrons, and another seven squadrons with MiG-19s and Sukhoi Su-7B ground-attack aircraft. There are Antonov An-2, Ilyushin Il-14M and Tupolev Tu-104 transports; Mil Mi-1 and Mi-4 helicopters; and Yakovlev Yak-11 and Yak-18, Zlin 226, MiG-15UTI and L-29 Delfin trainers.

GHANA

Ghana Air Force

The Ghana Air Force was formed in 1959 with Indian and Israeli assistance, mainly in the form of instructors, although India also provided Hindustan HT-2 trainers, and a Piper Super Cub was also obtained. In 1960, RAF personnel were seconded to the GhAF, and de Havilland Canada Chipmunk trainers were also supplied. Late in 1960 the first of a number of de Havilland Canada transport aircraft, DHC-2 Beavers, were obtained, and since then these have been joined by DHC-3 Otters and DHC-4 Caribou.

In 1963 the Nkrumah regime obtained from the USSR a number of aircraft, including Mil Mi-4 helicopters, and Antonov An-12 and Ilyushin Il-18 transport aircraft, which were rarely used and were returned to the Soviet Union when President Nkrumah was deposed. A few Aermacchi MB.326F jet trainers were ordered during the late 1960s.

Currently the GhAF has just over 1,000 men, one squadron of de Havilland Canada DHC-2 Beavers, of which there are twelve, one squadron of eleven DHC-3 Otters, and a third squadron with eight DHC-4 Caribou transports. A helicopter squadron operates six Westland S.55 Whirlwind, three Westland S.58 Wessex and four Hughes 269 aircraft, while there are also three de Havilland Herons, five Aermacchi MB.326F trainers, and a number of de Havilland Canada Chipmunk and Hindustan HT-2 trainers.

GREECE

Royal Hellenic Air Force

Greece occupies an important strategic position which offers her control over the eastern Mediterranean, and for which air power is, of course, invaluable. The Royal Hellenic Army established an air squadron with four Farman biplanes in 1912, and this saw service during the Balkan Wars of 1912–13. In 1914 the Royal Hellenic Navy formed an aviation service using Farman and Sopwith seaplanes with Royal Navy assistance. The two air arms were amalgamated to form the Hellenic Air Service in 1916, and this undertook operations against German and Turkish bases in Bulgaria before re-dividing itself the following year.

After the war ended, both the Royal Hellenic Naval Air Service and the Royal Hellenic Army Air Force, as they had become, were strengthened using an assortment of British and French aircraft respectively. However, in 1924 Greece was declared a republic and the 'Royal' prefixes were dropped. During the severe budgetary difficulties which beset Greece during the late 1920s, a Greek Air Force Fund was founded which was endowed with property and given fund-raising authority; several towns raised funds for the purchase of aircraft, notably Salonika which provided the price of twenty-five aircraft. The HNAS received Avro 504N trainers, Hawker Horsley torpedo-bombers, Greek-built Blackburn Velos (Dart) torpedo and reconnaissance-seaplanes and Armstrong-Whitworth Atlas general-purpose biplanes. The HAAF continued its preference for French equipment, buying Morane-Saulnier trainers and Breguet Br.19A and Br.19B reconnaissance-bombers.

A unified, autonomous air force came into existence in 1931 with the founding of the Hellenic Air Force, which gained its 'Royal' prefix with the restoration of the monarchy in 1935. A further assortment of aircraft were added to the RHAF's strength before the successful Axis invasion of 1941, including PZL P-24 and Bloch M.B.151 fighters; Bristol Blenheim, Fairey Battle and Potez 63 bombers; Henschel Hs 126 army co-operation aircraft; Avro Anson reconnaissance-bombers; and Dornier Do 22 and Fairey IIIF seaplanes.

The RHAF managed to ward off the Italian attack in 1940, but a shortage of suitable aircraft and experienced pilots left it unable to withstand the German invasion in 1941. Nevertheless, some personnel managed to escape to Egypt where fighter squadrons were formed using Hawker Hurricanes, and a bomber squadron used Bristol Blenheims; towards the end of the war these were replaced with Supermarine Spitfires and Martin Baltimores respectively.

For five years after liberation in 1944 there was civil war in Greece as Communists tried to seize power. During this period the RHAF was employed on police and counter-insurgency duties with its wartime equipment, plus Avro Ansons, Airspeed Oxfords and Douglas C-47 Dakotas, Curtiss SB2C-4 Helldiver dive-bombers, de Havilland Tiger Moth and North American Harvard trainers, and Convair-Stinson L-5 liaison aircraft.

The end of the civil war saw the RHAF in a much weakened condition, and when Greece joined the North Atlantic Treaty Organization in 1952 she became eligible for American assistance under the Mutual Defence Aid Programme. Personnel were trained in the United States and Germany to provide a sufficient number of pilots to form a basis for expansion, and eighty Canadair F-86D Sabre Mk. 2 and Mk. 4 fighters, 250 Republic F-84G Thunderjet fighter-bombers and RF-84F Thunderflash reconnaissance-fighters, and a number of Canadair T-33A jet trainers were delivered. Additional aircraft which followed during the following few years included additional Douglas C-47 transports, Greece's first helicopters, Sikorsky H-19D Chickasaws, and North American T-6G Texan trainers. During the late 1960s, American arms supplies were temporarily witheld, but restarted in 1970 with deliveries of Convair F-102 Delta Dagger interceptors.

Currently the Royal Hellenic Air Force, which is largely assigned to the Sixth Allied Tactical Air Force, has some 23,000 men, and operates four Northrop F-5A Freedom Fighter squadrons, four Northrop F-84F Thunderstreak and one RF-84F Thunderflash squadrons, and one Convair F-102 Delta Dagger, and two Lockheed F-104G Starfighter squadrons, with eighteen aircraft per squadron. There is one squadron of Grumman HU-16 Albatross amphibians, thirty Douglas C-47 Dakota and numbers of Nord Noratlas and Fairchild C-119G Packet transport aircraft; one squadron with twelve Sikorsky H-19 and six

Agusta-Bell 205 helicopters; and another squadron with ten Bell 47G Sioux helicopters; Cessna T-37B and T-41, Lockheed T-33A, Northrop F-5B and Lockheed TF-104G Starfighter trainers. Nike-Ajax and Nike-Hercules surface-to-air missiles are also employed.

GUATEMALA

Guatemalan Air Force   *Fuerza Aérea de Guatemala*

Guatemalan military aviation started with the visit of a French Aviation Mission in 1919, following which the Guatemalan Army formed an air unit using Avro 504 trainers. In 1929 the Cuerpo de Aeronáutica Militar was formed to operate a miscellany of French aircraft, until 1937 when Boeing P-26A fighters, Waco general-purpose aircraft and Ryan STM-2 trainers were placed in service. An agreement allowing Guatemalan bases to be used by the Allies during World War II resulted in deliveries of Douglas C-47 transports, Boeing-Stearman PT-17 Kaydet, North American AT-6 and Vultee BT-15 trainers. The arrival of an American Aviation Mission in 1945 resulted in the FAG receiving a squadron of North American F-51D Mustang fighters. Although Guatemala joined the Organization of American States in 1948, it was not until 1960 that ten Douglas B-26 Invader bombers and some Lockheed T-33A jet trainers were delivered.

Currently the FAG has 1,000 men, and operates eleven North American F-51D Mustang fighter-bombers, five Douglas B-26 Invader bombers, six Douglas C-47 Dakota transports, a Hiller UH-12B helicopter, and North American T-6 and Lockheed T-33A trainers.

GUINEA REPUBLIC

Guinea Air Force   *Force Aérienne de Guinea*

French Guinea became the Guinea Republic in 1958, and afterwards left the French Community. A small air force exists and, with the departure of President Nkrumah of Ghana to

Guinea, Russian military aid has been promised. A Bell 47G Sioux helicopter is operated, and Mikoyan MiG-17 fighters have been offered by the USSR, although it is not known whether any have actually been delivered.

HAITI

Haitian Air Corps

The Haitian Air Corps was founded in 1943, with American assistance, as a part of the Haitian Army, with the object of carrying mail between the main towns. After World War II ended, and Haiti became a member of the Organization of American States in 1948, the United States supplied North American F-51D Mustang fighters, and Beech, Boeing, Fairchild and Vultee trainers. Douglas C-47 Dakota transports were added to the fleet at a later date.

Current strength is 250 men, and includes six North American F-51D Mustang fighter-bombers; two Beech C-45 and three Douglas C-47 Dakota transports; two North American T-6G Texan and two T-28A Trojan trainers.

HONDURAS

Honduras Air Force *Fuerza Aérea Hondurena*

After the end of World War I, the Honduras Army formed an Air Corps using two Bristol F.2B fighter biplanes for AOP and liaison duties. Little further progress was made before World War II started, although some Boeing 95 mailplanes had been used as light bombers. During the war years, in return for support in safeguarding the Panama Canal, a variety of Beech, Boeing, Fairchild, North American, Ryan, Vultee and Waco trainers were supplied by the United States, plus Beech C-17 and Noorduyn Norseman transports.

In 1948, Honduras became a member of the Organization of American States, and the American military aid which followed included Lockheed P-38 Lightning and Bell P-63 Kingcobra fighters, followed by North American F-51D Mustang and Republic F-47D Thunderbolt fighter-bombers, and Beech C-45 and Douglas C-47 Dakota transports. The fighter-bombers were later replaced by Chance-Vought F4U Corsair fighter-bombers in 1958, plus a few Douglas C-54 transports, Lockheed T-33A jet trainers, and Cessna liaison aircraft. Three Sikorsky H-19 helicopters were also purchased.

Current equipment includes twelve Chance-Vought Corsair F4U fighter-bombers; four Douglas C-47 Dakota and a C-54 transports; three Sikorsky H-19 helicopters; some Cessna 180 and 185 liaison aircraft; and Lockheed T-33A, Beech AT-11 and North American T-6G Texan trainers.

HONG KONG

Hong Kong Auxiliary Air Force

Three Auster AOP 6 aircraft and two Sud Alouette III helicopters are operated by part-time local personnel under RAF supervision.

HUNGARY

Hungarian Air Force

Formerly a part of the Austro-Hungarian Empire, and therefore joined with the Central Powers in World War I, Hungary became an independent republic in 1918 but was not permitted to possess an air force under the Treaty of Versailles. In 1936 however, a small military air arm was formed using Fiat C.R.32 fighters, Meridionali Ro.37 reconnaissance and Heinkel He 46 AOP aircraft, with Hungarian-designed Weiss W.M.13 trainers. During 1938, German assistance was provided, including instructors and advisers and also Junkers Ju 86D and Heinkel He 70 bombers, and Bücker Bü 131 Jungmann trainers. The following year Fiat C.R.42 fighters, Caproni Ca.135 and C.310 bombers and Nardi trainers were added to the growing strength of what had become the Hungarian Army Air Force. Hungary entered World War II on the side of the Axis Powers in 1940.

During the war Hungarian units were sent to the Russian front. German aircraft, including Messerschmitt Bf 109 series fighters and Junkers Ju 87D Stuka dive-bombers and Ju 88A bombers, plus Italian Reggiane Re.2000 fighters, entered service in considerable numbers, a fair proportion of these being built in Hungary. Later, Messerschmitt Me 210 fighter-bombers entered Hungarian Army Air Force service, and Hungary even supplied these aircraft to the Luftwaffe. Hungary and its armed services suffered badly during the war and surrendered early in 1945.

After the war Hungary was forbidden to operate more than a small fighter and transport force, but in 1949 a Communist government took control of the country and Russian military aid commenced. Initially Yakovlev Yak-9 fighters, Lisunov Li-2 (C-47) transports and a variety of trainers were supplied. During the early 1950s the first jets, Mikoyan MiG-15 fighters, were supplied, along with obsolete Ilyushin Il-10 ground-attack aircraft and Tupolev Tu-2 bombers; and followed by MiG-17 fighters and Ilyushin Il-28 jet bombers; Antonov An-2 transports; and additional Li-2s; Mil Mi-1 and Mi-4 helicopters; Yakovlev Yak-11 and Yak-18 trainers; and MiG-15UTI jet trainers. Hungary became a member of the Warsaw Pact in 1955, but the following year a revolution brought Hungarian and Russian forces into opposition, and after Russia had quelled the revolution the Hungarian Air Force was stood down for some time. It is only in recent years that the Hungian Air Force has been modernized and again treated as an effective ally by the Soviet Union.

Currently the Hungarian Air Force has some ten or twelve squadrons operating MiG-21 interceptors, MiG-19 fighter-bombers, and MiG-17 and Sukhoi Su-7B ground-attack aircraft; with one Ilyushin Il-28 bomber squadron; Ilyushin Il-14, Antonov An-2 and Lisunov Li-2 transport aircraft; ten Mil Mi-1 and Mi-4 helicopters; and Yak-11 and Yak-18, L-29 Delfin and MiG-15UTI trainers. There are also some SA-2 Guideline surface-to-air missiles.

INDIA

Indian Air Force

India's history of military aviation dates from 1933 with the formation of the Indian Air Force under RAF control, initially to operate four Westland Wapiti general-purpose aircraft on army co-operation duties. It was not until 1940 that a complete squadron of Wapitis was operated, but considerable expansion took place during the years which followed, starting with the replacing of the Wapitis by Westland Lysanders in 1941, By 1943 the IAF had seven Hawker Hurricane fighter squadrons and two Vultee Vengeance dive-bomber squadrons. Pilots were trained mainly in Australia and Canada, although some were trained in India and the United Kingdom. In 1944, Supermarine Spitfire fighters started to arrive. Mainly the IAF operated in support of Allied army units.

The 'Royal' prefix was granted in 1945 by King George VI in recognition of the IAF's wartime achievements, and a squadron was sent to Japan as a part of the Commonwealth Occupation Force. The division of India into two countries, India and Pakistan, on being granted independence in 1947, also saw the RIAF split into two. India retained seven fighter squadrons operating Hawker Hurricanes and Tempests and Supermarine Spitfires, plus a Douglas C-47 Dakota transport squadron. Immediately after independence, additional aircraft were delivered, mainly Tempests but with some Spitfires, while Hindustan Aircraft Industries refurbished many redundant Consolidated B-24 Liberator bombers to provide a small bomber force. De Havilland Vampire F.3 jet fighters and FB.9 fighter-bombers were also bought at this time, along with de Havilland Devon transports; while Percival Prentice trainers were built in India pending production of the nationally-designed Hindustan HT-2 trainer. The 'Royal' prefix was dropped in 1950 when India became a republic within the British Commonwealth.

More than one hundred Dassault Ouragan jet fighter-bombers were bought in 1953, followed in 1954 by twenty-six Fairchild C-119G Packet transports, and in 1955 by Auster AOP 9 aircraft, and Vickers Viscount 700 and Ilyushin Il-14 transports. In 1956, more than one hundred Dassault Mystère IVA fighters were ordered, while licence-production of the Folland Gnat fighter started. English Electric Canberra B.8 jet bombers, Hawker Hunter F.56 fighters and T.7 trainers, licence-built de Havilland Vampire T.55 trainers, and de Havilland Canada DHC-3 Otter transports, Bell 47G Sioux and Sikorsky S.55 helicopters were all delivered during the late 1950s, usually in quite large numbers, for the rapidly expanding IAF. North American T-6G Texan trainers were also produced under licence.

The 1960s saw licence-production of the Mikoyan MiG-21 interceptor, while the nationally-designed Hindustan HK-24 Marut also entered production. Conflicts with Pakistan over border disputes proved the Gnat to be a particularly successful design, and this aircraft was put back into production during the decade. Amongst the other aircraft of the period can be included Antonov An-12 transports, Mil Mi-4 helicopters, and de Havilland Canada DHC-4 Caribou transports. There were also several border clashes with Communist China during this period.

Currently the IAF has 90,000 men, and operates six Mikoyan MiG-21 interceptor squadrons; eight Hawker Siddeley Gnat fighter squadrons; two Hindustan HF-24 Marut; four Sukhoi Su-7B and six Hawker Hunter fighter-bomber squadrons; one Canberra PR.57 reconnaissance and three BAC Canberra B.8 bomber squadrons; two de Havilland Vampire tactical reconnaissance squadrons; one Lockheed L-1049 Super Constellation maritime-reconnaissance squadron; sixty Douglas C-47, sixty Fairchild C-119G Packet, twenty-four Ilyushin Il-14, thirty Antonov An-12, thirty de Havilland Canada DHC-3 Otter and fifteen DHC-4 Caribou, and twenty-five Hawker Siddeley 748 transport aircraft; one hundred Mil Mi-4 and 120 Sud Alouette III helicopters, with a few Bell 47G Sioux still remaining in service; along with a number of Sikorsky types; Hawker Hunter T.7, de Havilland Vampire T.55, North American T-6G Texan, Hindustan HT-2 and HT-16, and MiG-21UTI trainers. About fifty Krishaks are operated on AOP duties for the Army.

Indian Naval Aviation

In 1950 a Fleet Requirements Unit was established as a first move towards forming a fleet air arm, and this was equipped with Short Sealand amphibians, five Fairey Firefly target-tugs, and three Hindustan HT-2 trainers. The former HMS *Hercules* was refitted and joined the Indian Fleet in 1961 as the INS *Vikrant*, India's first, and so far only, aircraft carrier. Hawker Sea Hawk attack aircraft and Breguet Br.1050 Alizé anti-submarine aircraft were obtained for carrier-borne operations, while a further fifteen Hindustan HT-2 and jet de Havilland Vampire T.55 trainers were also bought.

Currently the force consists of the 16,000-ton aircraft carrier, INS *Vikrant*, a total of thirty-five Hawker Sea Hawks in two squadrons, twelve Breguet Br.1050 Alizé aircraft, six Westland SH-3D Sea King and ten Sud Alouette III helicopters, plus the trainers already mentioned. The INS *Vikrant* carries sixteen Sea Hawks, four Alizé and two Alouette III helicopters at any one time.

INDONESIA

Indonesian Republican Air Force
Angatan Udera Republik Indonesia

Although the formal grant of independence to Indonesia did not take place until December 1949, an aviation division of the People's Peace Preservation Force was established in December 1945, during the interim period after the Japanese surrender and before the return of Dutch forces in the spring of 1946. Initially there were only five Indonesian pilots and a few ex-Japanese aircraft, which were in poor condition after Dutch air attack. Indian assistance was provided just before and after independence, and Dutch assistance was also provided for the Angatan Udera Republik Indonesia (AURI) after independence. Ex-RNAF aircraft provided for AURI after independence included North American F-51D and F-51K Mustang fighter-bombers and B-25 Mitchell bombers; Convair PBY-5A Catalina amphibians; Lockheed 12A transports, used mainly for domestic air services;

Piper L-4J Cubs, Auster IIIs and Aiglets for training and AOP duties. India provided Hindustan HT-2 trainers.

IAF personnel were seconded to Indonesia in 1955 to reorganize the AURI, while at this time pilots were being trained in the United States, the United Kingdom and the Netherlands. Eight de Havilland Vampire fighter-bombers were obtained from Great Britain in 1955, while in 1958 deliveries of aircraft from Czechoslovakia commenced, including about forty Ilyushin Il-28 light bombers and sixty Mikoyan MiG-17 fighters. Training of pilots for these aircraft took place in Egypt. During the late 1950s further Western types were delivered, including de Havilland Canada DHC-3 Otters, Cessna 180s, Grumman HU-16 amphibians, Fairey Gannet AS.4 maritime-reconnaissance aircraft, with a number of T.5 trainers, Lockheed C-130B Hercules transports, and Sikorsky S-58 helicopters.

An aggressive foreign policy, first manifesting itself in an invasion of West Irian in 1962, brought Indonesian and Dutch forces into conflict and, followed by a confrontation with Great Britain, Australia, New Zealand and Malaysia over the creation of the Federation of Malaysia, cut the AURI off from Western arms supplies. This period, which lasted until 1966, saw Mikoyan MiG-19 fighters, Tupolev Tu-16 bombers and Ilyushin Il-14 and Antonov An-2 transports enter service, followed by a few MiG-21 interceptors. AURI's operations in support of the policy of confrontation were largely unsuccessful, and this period ended with the stripping of power from the then President, Sukarno.

Currently the AURI has some 50,000 men, although this number includes some paratroops, and operates fifteen Mikoyan MiG-21 interceptors; thirty-five MiG-19, forty MiG-17 and twenty MiG-15 fighter-bombers, with a few North American F-51D Mustangs also still operating in this role; there are thirty Ilyushin Il-28, twenty-five Tupolev Tu-16, and fifteen North American B-25 bombers; some fifty or sixty transports, including Lockheed C-130B Hercules, Fokker F-27 Troopship, Douglas C-47 Dakota, Ilyushin Il-14, Antonov An-12, Short Skyvan 3M, and de Havilland Canada DHC-3 Otter; twenty Mil Mi-4 and six Mi-6, two Bell 204B, some Sikorsky S-61A and seven Sud Alouette III helicopters, and an assortment of training and liaison aircraft.

Indonesian Naval Aviation *Angatan Laut Republik Indonesia*

Currently the Indonesian Navy operates a total of twenty Mikoyan MiG-19 and MiG-21 fighters and interceptors, five Grumman HU-16 Albatross and some Consolidated PBY-5A Catalina amphibians, originally delivered to AURI, as were the Fairey Gannets of which less than a handful survive. There are a number of Sikorsky S-58, four Bell 47G Sioux and three Sud Alouette II helicopters.

IRAN

Imperial Iranian Air Force

Currently one of the most modern air forces in the Middle East, the history of the Imperial Iranian Air Force dates from the founding of an Air Office of the Iranian Army in 1922, accompanied by the purchase of a Junkers F 13 and the hire of a German pilot. The following year an Aero A.30 was bought for AOP duties, and in 1924 four D.H.4s and D.H.9s and further F 13s were added to the small air arm. The first Iranian pilots went to France for training in 1924, and in the summer of that year the Air Office became the Iranian Air Force, although still a part of the Army.

Formation of the Imperial Iranian Air Force in 1932 as a separate service was accompanied by a modernization programme, including the purchase of eighteen de Havilland Tiger Moth trainers in 1932, and twelve Hawker Fury fighters in 1933; and these were followed by Hawker Hart light bombers and thirty-two Audax AOP aircraft. RSwAF personnel were seconded to the IIAF during the 1930s as advisers and instructors.

In 1938 thirty-eight Hawker Hurricane and a number of Curtiss 75A Hawk fighter-bombers were ordered, but only two Hurricanes were delivered before World War II prevented further deliveries. Iran was invaded by British and Russian forces in 1941 to guarantee a supply route to Russia and prevent Iran from joining the Axis powers. During the war the IIAF largely stagnated, except that some Avro Ansons and de Havilland Tiger Moths were delivered in 1943 after Iran declared war on Germany.

After the war the IIAF returned to a more active existence, with the delivery of thirty-four Hawker Hurricane fighters, and a number of American aircraft, including Republic F-47D Thunderbolt fighter-bombers, Douglas C-47 Dakota transports, Piper L-4 AOP aircraft, and North American T-6G Texan and Boeing-Stearman PT-13 Kaydet trainers. During the late 1940s all IIAF aircrew were trained in the United States and Germany, but by 1950 it was possible to resume training in Iran. American military aid during the 1950s included Republic F-84F Thunderstreak and North American F-86F Sabre fighter-bombers, and Lockheed T-33A trainers. In recent years these aircraft have been supplemented and to some extent replaced by Northrop F-5As and McDonnell Douglas F-4D Phantom IIs. Iran was a founder-member of the Baghdad Pact, which since Iraq's withdrawal has been renamed the Central Treaty Organization.

Current strength of the IIAF is 17,000 men, and there are two squadrons with a total of thirty-two McDonnell Douglas F-4D Phantom II and five squadrons with a total of eighty Northrop F-5A fighter-bombers; while there are two Northrop RF-5A tactical reconnaissance squadrons, and a number of North American F-86 Sabres operating in the interceptor role; ten Lockheed C-130E Hercules, eight Beech C-45, ten Douglas C-47, and five de Havilland Canada DHC-2 Beaver transports; one hundred Agusta-Bell 206A, forty 205, forty Bell UH-1D, Kaman HH-43B Huskie, sixteen Sud Super Frelon and sixteen Italian-built Boeing-Vertol CH-47 Chinook helicopters; Lockheed T-33A, Northrop F-5B and North American T-6G Texan trainers. Short Tigercat surface-to-air missiles are also deployed. A number of these helicopters are operated on behalf of the Imperial Iranian Army, while the Imperial Iranian Gendarmerie operates five Agusta-Bell 205 helicopters.

IRAQ

Iraqi Air Force

The Iraqi Air Force was formed in 1931 as the Royal Iraqi Air Force when the first Iraqi pilots returned from training in the

United Kingdom with five de Havilland Gipsy Moth trainers, which were soon followed by another four, and in 1932 by four Puss Moths fitted with bomb racks. Immediately following its formation, the RIAF was used to suppress tribal uprisings. In 1934, Hawker Nisrs, de Havilland Dragons and Tiger Moths, and more Puss Moths were delivered. Savoia-Marchetti S.M.79B bombers and Breda Ba.65 fighter-bombers arrived in 1937, while Gloster Gladiator fighters, Avro Anson and Douglas DB-8A bombers, and de Havilland Dragon Rapide and Dragonfly light transports were delivered in 1938–40.

A German-inspired uprising in 1941 brought the RIAF into conflict with the RAF, with the loss of most of the former's aircraft excepting a few Gladiators, Nisrs and Tiger Moths. The RIAF remained inactive for the rest of the war.

Re-equipment began in 1946 with thirty Hawker Fury fighter-bombers, and in 1948 four Bristol Freighters, some de Havilland Doves, and Auster AOP 6 and T.7 aircraft were delivered. As a founder-member of the Baghdad Pact along with Great Britain, Turkey, Iran and Pakistan, Iraq was able to obtain more British equipment throughout the early and mid-1950s. This included twenty de Havilland Canada Chipmunk T.20 trainers in 1951; twelve de Havilland Vampire FB.52 fighter-bombers and six T.55 trainers in 1953 (these were Iraq's first jets); Iraq's first helicopters, two Westland S.51 Dragonflies in 1955; along with two de Havilland Venom FB.50 fighter-bombers, two de Havilland Heron light transports in 1956; and in 1957 a squadron of Hawker Hunter F.6 fighters.

Revolution in 1958, with the assassination of the royal family, led to the founding of a republic, and meant the dropping of the 'Royal' prefix from the Iraqi Air Force's title. Iraq left the Baghdad Pact, which was renamed the Central Treaty Organization, and RAF assistance, which had been given for some years, ceased with the start of Russian military aid. October 1958 saw the arrival of twenty Mikoyan MiG-15 fighter-bombers, and these were followed by Russian instructors and advisers with additional MiG-15s and Ilyushin Il-28 jet bombers. This trend continued throughout most of the 1960s with deliveries of MiG-17, MiG-19 and MiG-21 fighters and interceptors, Antonov An-12 transports, and Mil Mi-1 and Mi-4 helicopters. In 1965 and 1966 a series of defections of MiG-21 pilots led to the suspension of Russian supplies, and this with a serious spares shortage led to the purchase of a number of British aircraft, including Hawker Hunter F.G.A.9s, F.R.10s, and T.66/9s, Westland S.58 Wessex helicopters and BAC Jet Provost T.52 jet trainers.

Following the Arab-Israeli War of 1967, Russian arms supplies recommenced towards the end of the 1960s.

Currently the IAF has some 7,500 men and operates sixty Mikoyan MiG-21 interceptors; fifty MiG-19 and MiG-17 fighter-bombers; thirty-six Hawker Hunter Mk. 9 and fifty Sukhoi Su-7B ground-attack aircraft; ten Ilyushin Il-28 and eight Tupolev Tu-16 bombers; twelve Antonov An-2, six An-12 and ten An-24, thirteen Ilyushin Il-14, two de Havilland Heron and three Bristol 170 Mk. 31 Freighter transport aircraft; four Mil Mi-1 and twenty Mi-4, and eleven Westland S.58 Wessex helicopters; twenty BAC Jet Provost T.52 trainers, with possibly some MiG-15UTI, Yakovlev Yak-11 and Yak-18 and L-29 Delfins also in this role. SA-2 'Guideline' surface-to-air missiles are also deployed.

ISRAEL

Israel Defence Force/Air Force

As its name implies, the Israel Defence Force/Air Force is a part of a unified defence service. Its origins date from before the founding of the modern state of Israel, when the Zionist Haganah underground movement formed the Sherut Avir which undertook AOP duties using Auster and Taylorcraft light aircraft for the Haganah movement prior to the termination of Britain's mandate in Palestine. These first aircraft, and also those of the Chel Ha'avir (as it became after independence in 1948), were usually built from cannibalized aircraft.

An attack by Egyptian Spitfires in 1949 prompted the Chel Ha'avir to equip itself with more potent aircraft, at first by cannibalizing former British Supermarine Spitfire fighters and de Havilland Mosquito fighter-bombers left behind on former RAF bases after withdrawal, but later some three hundred aircraft of these two types were obtained second-hand, plus a few Avia C.210

fighters. Before the formation of the Israel Defence Force/Air Force in 1951, Boeing B-17G Fortress bombers, Curtiss C-46, Douglas C-47 and C-54 transports, and Boeing-Stearman PT-17 Kaydet, North American T-6 Harvard and Avro Anson and Airspeed Oxford trainers were introduced, and were followed immediately before the IDF/AF's formation by twenty-five ex-RSwAF North American F-51D Mustang fighter-bombers. Some of the transports were converted to bombers for raids on Cairo, while the Harvards were adapted for ground-attack duties. IDF/AF personnel originated from a wide variety of nationalities, in common with the rest of the population of Israel.

The first jet equipment, fourteen Gloster Meteor F.8 fighters, arrived in 1953, and were soon afterwards joined by six Meteor NF.13 all-weather fighters and T.7 trainers. In 1955 thirty new Dassault Ouragan fighter-bombers were delivered, and were soon supplemented by forty-five ex-Armée de l'Air Ouragans; while forty-one Fokker S-11 Instructors and eight Nord 2501 Noratlas transports were also put into service. In order to obtain a Middle East balance of power, Israel countered the delivery of Czech equipment to Egypt in 1955 by ordering twenty-four Canadair CL-13 Sabre 6 and twenty-four Dassault Mystère fighters – the order for the latter was increased to sixty when the Sabres were embargoed. During the Suez Crisis of 1956 the IDF/AF gained air superiority over the Egyptian Air Force with the loss of eleven aircraft, although the Egyptians had been numerically superior.

Since their water-cooled engines made them vulnerable to ground fire, the Mustangs were replaced by licence-built Potez Magister armed-trainers, during the late 1950s. This period also saw the arrival of twenty-four Sud Vautour II light bombers and a number of Dassault Super Mystère interceptors. The 1960s saw some sixty Dassault Mirage IIICJ fighter-bombers enter service, and this aircraft proved highly successful in the June 1967 war against the Arab countries, when the IDF/AF virtually wiped out the Egyptian and Jordanian Air Forces. A further fifty Mirages were ordered and built, but remain undelivered due to an embargo placed on these aircraft by the French Government. Israel now looks mainly to the United States for arms supplies in order to maintain a balance of power with the Arab nations. However, some Israeli construction of aircraft is also likely, and an Israeli-built Mirage development has flown.

Currently the IDF/AF has a total personnel strength of 17,000 men, and operates fifty McDonnell Douglas F-4E Phantom II fighter-bombers and six RF-4Es, sixty Dassault Mirage IIICJ, thirty Dassault Mystère and thirty Ouragan, and sixty-seven McDonnell Douglas A-4E and eighteen A-4M Skyhawk fighter-bombers. Ten or twelve Super Mystère interceptors remain in service. There are eighty-five Potez Magister armed-trainers; fifteen Nord Noratlas, and ten Douglas C-47 Dakota transports; six Boeing Stratocruiser tanker aircraft; twenty-five Agusta-Bell 205, twenty Sud Alouette III and twelve Super Frelon, eight Sikorsky CH-53 Sea Stallion and fifteen H-34 Choctaw helicopters; and Hawk surface-to-air missiles. A further sixteen Phantom IIs and fifty refurbished ex-USN LTV F-8 Crusaders may be delivered in the near future.

Italian Air Force *Aeronautica Militare Italiano*

Italy's first military aircraft were bought in 1911 during the Italo-Turkish War when the Army formed an aeroplane company using Blériot XIs, Etrich Taubes, Maurice Farman S-11s and Nieuports on reconnaissance duties. The following year, the Battagliore Aviatori (Aviation Battalion) was formed, along with the Servizio d'Aviazione Coloniale (Colonial Air Service), although, by the end of the year, the Battagliore Aviatori had become the Military Aviation Service. Yet another change of title occurred in 1914 when the MAS became the Corpo Aeronautico Militare (Military Aviation Corps).

Italy fought on the side of the Allies during World War I, during which the CAM's strength rose from seventy aircraft of Blériot, Nieuport, Maurice Farman and Caproni manufacture, to 1,800 aircraft, including Nieuport 17C-1 Bébé and 110, Spad S.VII, Hanriot HD-1 and Macchi M.14 fighters; Caproni Ca.33, Ca.40 and Ca.46 and Macchi M.7 and M.8 bombers; and reconnaissance types such as the Ansaldo S.V.A.4, S.V.A.5, S.V.A.9 and S.V.A.10, Savoia-Pomilio S.P.3 and S.P.4 and the Fiat R.2.

Naturally, the strength of the CAM dropped sharply after the

Armistice, and, in common with many air forces and air arms, it suffered from considerable neglect until Benito Mussolini assumed power in 1923. That same year the CAM was reorganized as a separate service, the Regia Aeronautica. Steps were taken to strengthen the new force, and to boost morale – one means of achieving the latter was the organizing of mass long-distance flights such as the one in which twenty-four Savoia-Marchetti S.M.55X flying-boats flew from Rome to New York and back in 1933. The RA also had 1,200 aircraft by 1933, divided amongst thirty-seven fighter squadrons equipped with Fiat C.R.20s and C.R.30s; thirty-four bomber squadrons equipped with Caproni Ca.73s and Ca.101s; thirty-seven reconnaissance squadrons equipped with Romeo Ro.1s, Caproni Ca.97s and Fiat R.22s; and a number of seaplane and flying-boat squadrons mainly equipped with Savoia-Marchetti S.M.55Xs; with transport squadrons equipped with Caproni Ca.101s, Ca.111s and Ca.133s, and Savoia-Marchetti S.M.81s.

Italian forces invaded Ethiopia in October 1935, and the RA played an important role in the operation although most of the aircraft deployed were transports since there was no Ethiopian air defence of any consequence. In 1936 the Spanish Civil War started, and strong RA contingents were sent to fight alongside the German forces.

Italy entered World War II as an ally of Germany in 1940, and by which time the RA had some 3,000 aircraft, including some four hundred obsolete and obsolescent types serving in the African colonies. The RA was operational mainly in the Mediterranean area, especially in Greece, North Africa and over Malta, but a token force of seventy-five Fiat B.R.20M Cicogna bombers, and fifty C.R.42 and G.50 fighters were based in Belgium and flew sorties over the British Isles. A force sent to the Russian front fared badly. During the early part of the war the first German aircraft entered RA service, notably Junkers Ju 87 Stuka dive-bombers, while some Daimler-Benz DB 601 and DB 605 engines were supplied to the Italian aircraft industry, which had been without efficient low-drag air-cooled engines for its fighters. Several Italian designs, including the Fiat G.52 and G.55 fighters, became exceptionally good aircraft when fitted with German engines. As the war progressed, more German types entered RA service, including the Messerschmitt Bf 109F and Bf 110G fighters and Dornier Do 217 bombers.

On 8 September 1943, Italy capitulated, and many of the RA's surviving squadrons flew south to join the Allies, forming the Italian Co-Belligerent Air Force, while that part of Italy under German occupation became the Republica Sociale Italiana, and its air force, the Aviazone della RSI.

After the war ended, the Aeronautica Militare Italiano was formed with a variety of ex-wartime Italian aircraft plus a number of Supermarine Spitfire and Bell P-39Q Airacobra fighters, Martin Baltimore bombers, and Douglas C-47 Dakota transports. Under the terms of the 1947 Peace Treaty, only two hundred out of a permitted 350 aircraft could be combat types. These restrictions were removed when Italy became a member of the North Atlantic Treaty Organization in 1949.

During the early 1950s a major re-equipment programme was set in motion. The first aircraft were eighty Lockheed P-38J Lightning fighters, one hundred Beech C-45 transports, and a number of Stinson L-5 liaison aircraft, the gift of the United States Government, and they were supplemented by some Italian types, Fiat G.212 transports and G.46 and G.59 trainers; while in 1950 the first jets, de Havilland Vampire FB.5 fighter-bombers, were delivered, and both Fiat and Aermacchi obtained licences for production of this aircraft. For the most part, however, American equipment was introduced during this period, including Republic F-47D Thunderbolt and F-84F Thunderstreak fighter-bombers; and RF-84F Thunderflash reconnaissance-fighters; Curtiss SB2C-5 Helldiver and Lockheed PV-2 Harpoon maritime-reconnaissance aircraft; Grumman S2F-1 Tracker anti-submarine aircraft; North American T-6G Texan and Lockheed T-33A trainers; and Grumman SA-16A Albatross amphibians. Some de Havilland Vampire NF.54 and Canadair F-86 Sabres were introduced in 1955.

By 1960, Aermacchi MB.326 trainers had started to replace the Texan, and helicopters had been introduced, including the Sikorsky UH-19 and SH-34J, but mainly licence-built American Bell designs of which the first was the Agusta-Bell 47G Sioux. Since 1960, Lockheed F-104G Starfighters and Fiat G.91 fighters, with Fairchild C-119G Packet transports, have been amongst the types introduced.

Currently the Aeronautica Militare Italiano, which has 73,000 men, operates six Lockheed F-104G/S Starfighter, five Fiat G.91R-1, R-4 and G.91Y, two Republic F-84F Thunderstreak and two Canadair F-86K Sabre fighter-bomber squadrons; which are assigned to NATO (Fifth Allied Tactical Air Force); along with two Republic RF-84F Thunderflash reconnaissance-fighter squadrons, and three Fairchild C-119G Packet transport squadrons; while a further two Fiat G.91 fighter-bomber squadrons and a Douglas C-47 Dakota, a Convair 440 and Douglas DC-6 transport squadron are also operated; and there are also ninety Agusta-Bell 47G/J Sioux, sixty 204B and three 205, two Bell JetRanger, four Sikorsky UH-19, and twelve Agusta A101G helicopters. Three squadrons of Nike-Ajax and Nike-Hercules surface-to-air missiles are operated. Trainers include Aermacchi MB.326 and Savoia-Marchetti SF.260s.

Numbers of Fiat G.222 tactical transports may be introduced in the near future, while after 1975 one hundred Panavia 200 Panther multi-role combat aircraft will enter service as Starfighter replacements.

Italian Naval Aviation *Marinavia*

Formed since the end of World War II, the Marinavia currently operates twenty-four Sikorsky SH-3D and six SH-34J, forty Agusta-Bell 47G/J Sioux, twenty-four 204B and thirty Agusta A106 helicopters from shore bases and Italian warships. Experiments have taken place in operating Hawker Siddeley Harrier vertical take-off strike aircraft from the landing platform of the cruiser *Andrea Doria*, and aircraft of this type might be purchased in the future.

Italian Army Aviation *CAALE*

The Italian Army operates twenty-six licence-built Boeing-Vertol CH-47C Chinook medium-lift helicopters; 125 Agusta-Bell 47G/J and seventy 204B helicopters on liaison, AOP and communications duties; plus 150 Piper L-18/21 and Cessna O-1E light aircraft which are gradually being replaced by Aerfer AM-3C and Savoia-Marchetti S.M.1019 liaison aircraft.

IVORY COAST

Ivory Coast Air Force   *Force Aérienne de Côte d'Ivoire*

A former French colony still maintaining close links with France, the Ivory Coast became independent in 1960. The air force, while small, is rather more ambitious than that of most former French colonies. Currently there are three hundred men, three Douglas C-47 Dakota and five Max Holste 1521M Broussard transports, with one Dassault Mystère 20, a Beech 18 and an Aero Commander 500 (the gift of the US Government) for VIP duties, with two Sud Alouette II, three Alouette III and two SA.330 Puma helicopters. Three Cessna 337s and a Fokker F-27 Troopship were obtained in 1971.

JAMAICA

Air Wing, Jamaica Defence Force

The Air Wing of the Jamaica Defence Force commenced operations in 1963 on liaison and light transport duties, for which it currently operates a de Havilland Canada DHC-6 Twin Otter and a Cessna 185, plus two Bell 47G Sioux helicopters.

JAPAN

Japanese Air Self-Defence Force

Japan's history of military aviation dates from 1911 when the Japanese Army Air Force and the Japanese Navy Air Force were both formed. Initial equipment for the JAAF included three Henri Farmans, two Wrights, an Antoinette and a Blériot, while the JNAF had two Maurice Farman and two Curtiss seaplanes. Starting in 1912, the JAAF received Japanese-designed Tokogawa 1 and Sei Model 1 and 2 biplanes, licence-built Maurice Farmans, and a Nieuport and a Rumpler Taube. And the JNAF received Otari and Ushioku biplanes, and a Blériot, followed after 1916 by

Short reconnaissance-seaplanes, Sopwith seaplane fighters, Déperdussin seaplane trainers, and Yokosuka 80 Model A seaplanes. Although Japan participated in World War I on the side of the Allies, this participation was largely limited to the occupation of the German Mandate Port at Tsingtao on the Chinese coast. A few Japanese pilots flew with the Aviation Militaire.

After the war, firm foundations were laid for the development of an aircraft industry, with the involvement of three industrial groups, Mitsubishi, Nakajima and Kawasaki, in aircraft development and manufacture. A French Aviation Mission visited Japan to advise on the structure and future of the JAAF, while a British Air Mission visited the JNAF.

During the early 1920s the JAAF was completely re-equipped with Spad S.13C and licence-built 20C, Nieuport 24C and 29C, plus fifty ex-RAF Sopwith 1½-Strutter and Pup fighters; Breguet Br.14B, Farman F.50 and licence-built F.60 Goliath bombers; Salmson SA-2 reconnaissance aircraft; Nieuport 81E, 83E and 24C, and Hanriot and Caudron C-6 trainers; plus the Nakajima Type 5 advanced trainer. A variety of aircraft types also entered JNAF service, including Sparrowhawk Mars IV and Mitsubishi Type 10 shipboard fighters, with reconnaissance and training versions of both; Short F.5 America, Schreck F.B.A.17 and Tellier flying-boats; Avro 504K and 504L trainers; plus some Sopwith, Airco, Vickers, Blackburn and Supermarine types. A Mars IV fighter was used for trials operating from a platform built over the forward gun turret of the battleship *Hamishiro* in 1922, and this led to ordering later the same year of the first Japanese aircraft carrier, the *Hosho*, which entered service equipped with Mars IV and Mitsubishi Type 10 fighters.

The late 1920s saw two more carriers, the *Akagi* and *Kaga*, enter service in 1928 and 1929 respectively. Nakajima A1N1 Type 3 (licence-built Gloster Gambet) fighters, Mitsubishi B2M1 Type 89 (adapted Blackburn type) naval bomber and the C1M2 reconnaissance aircraft were evolved for carrier-borne operations, with Yokosuka E1Y1 and Aichi Type 2 (licence-built Heinkel H.D.25) seaplanes, and Hiro H1H1, H1H2 and H2H1 flying-boats. During the early 1930s these aircraft were followed by Nakajima A2N1 shipboard fighters and advanced trainers, E4N1 and Kawanishi E5K1 reconnaissance seaplanes, and Hiro H3H1 flying-boats.

Meanwhile, the JAAF had equipped with Mitsubishi Type 87 and Kawasaki Type 87 (licence-built Dornier F) and Type 88 bombers, respectively light, heavy and reconnaissance. These were followed during the early 1930s by Nakajima Type 91 and Kawasaki Type 95 fighters, Mitsubishi Type 92 (licence-built Junkers G.38) and Type 93 and Kawasaki Type 93 bombers, and Nakajima Type 94 AOP aircraft.

In 1931, Japan invaded Manchuria in order to maintain a major export market, and Manchuria became a Japanese protectorate after a successful JAAF action. This was followed in 1932 by action against Shanghai, largely a JNAF attack from the carriers *Kaga* and *Hiro*, and afterwards a 'cold war' situation followed during which Japan sought to build up her air power. In 1932, two aircraft of the 7-Shi range of prototypes, the Hiro G2H1 Type 95 naval bomber and Kawanishi E7K1 Type 94 reconnaissance aircraft, were put into production. These were followed by the 9-Shi (ninth year of Showa reign) range of prototypes in 1934, including the Mitsubishi A5M1 Type 96 fighters and G3M1 Type 96 land-based long-range bomber and the Aichi D1A2 and Nakajima B4Y1 Type 96 carrier-borne bombers, the Watanabe E9W1 and Aichi E1A1 Type 96 reconnaissance seaplanes, and the Kawanishi H6K1 Type 97 four-engined flying-boat, which all entered service around 1936. These aircraft were followed by Nakajima B5N1 Type 97 carrier-borne strike aircraft, Mitsubishi C5M1 Type 98 reconnaissance aircraft, and Hiro H5Y1 Type 99 reconnaissance flying-boats.

Meanwhile the JAAF had also selected and introduced new aircraft: Kawasaki Type 95 and Nakajima Ki 27a and Ki 27b Type 97 fighters; Mitsubishi Ki 30 and Ki 21, and the Kawasaki Ki 48 Type 98 bombers; Nakajima Ki 34 Type 97 transport; the Tachikawa Ki 36 AOP aircraft; Mitsubishi Ki 15 and Ki 51 Type 97 reconnaissance aircraft; and Ki 51b ground-attack aircraft.

The conflict in China restarted in 1937, with the JAAF and JNAF outnumbering the Chinese forces by more than five to one. During the winter of 1937–8, Japanese forces advanced rapidly across China, although, in some areas outside of the range of fighter escorts, JAAF and JNAF bombers were subject to devasting Chinese fighter attack. New aircraft continued to enter service, and the most notable and useful of these was the Mitsubishi A6M2 Type 0 shipboard fighter – the Zero. New aircraft carriers had been put into service, including the *Akagi, Chitose, Chiyuda, Hiryu, Junyo, Ryujo, Shinano, Shokaku* and *Soryu, Taihu, Zuihu,* and *Zuikaku*. By the end of 1941, China was virtually defeated.

On 8 December 1941, aircraft from six Japanese carriers attacked the large United States naval base at Pearl Harbour, Hawaii. By all of the usual criteria the attack was a success, but since it brought the United States into World War II it could be construed as a major blunder. An attack on Hong Kong followed, and the island was occupied. Burma, Borneo, the Celebes, Rabaul, New Ireland, the Solomons, Sumatra and Java, the Philippines, Malaya and Singapore, and Thailand all followed, with JAAF and JNAF aircraft providing support wherever it was needed. The Vichy French Government had also allowed Japanese forces to use bases in French Indo-China. However, in 1942, American carrier-borne bombers attacked Tokyo and other major Japanese cities, and an attack by Japanese forces on the American-held island of Midway failed miserably, with four out of the six carriers used and most of the aircraft destroyed. The Imperial Japanese Navy was never really to recover from this defeat.

Major wartime types of the JNAF included developments of the very versatile Zero, including seaplane versions, Kawanishi H8K2 Type 2 flying-boats, Nakajima J1N1 Type 2 fighter-reconnaissance aircraft, and Aichi D4Y1 and D4Y2 and Mitsubishi G4M2 bombers. The JAAF, in common with the JNAF, used many pre-war designs and their developments, but new aircraft included the Nakajima Ki 44 Type 2 and Ki 84 Type 4, and Kawanishi Ki 4 and Ki 61 Type 3 fighters, Nakajima Ki 49 Type 0 and Mitsubishi Ki 67 Type 4 bombers and Kokusai Ki 76 AOP aircraft. The JAAF bombed India and Darwin, Australia.

Towards the end of the war, kamikaze (Divine Wind) suicide groups were formed, finding ready acceptance amongst JNAF and JAAF personnel since the Shinto philosophy dictates that death in battle is a sure way to Heaven. Initially, Zeros were used carrying 500 lb. bombs, but Oka piloted bombs were also dropped from heavy bombers, and a number of special aircraft types were developed for the JNAF, including the Aichi D4Y4, Mitsubishi D5Y1, the Nakajima Kitsuka and Yokosuka Oka – the last two being jets. The JAAF used the Nakijima Ki 115 and Ki 201.

However, in spite of the success of the kamikaze against American warships, including the loss of an aircraft carrier (no British carrier was lost although several were hit), the increased heavy bombing raids against Japan, the difficulties of intercepting the high-flying Boeing B-29 Superfortresses, and eventually the dropping of atomic bombs on Hiroshima and Nagasaki in August 1945, all led to Japanese surrender. Occupation by the Allies followed.

In 1950 a National Police Reserve Force was formed under American sponsorship, followed in 1954 by land, air and sea self-defence agencies.

Initially the Japanese Air Self-Defence Force, which was formed with American assistance, operated North American T-6G Texan and Beech T-34 Mentor trainers, with a few Lockheed T-33A jet and Curtiss C-46 Commando navigational trainers. The first students (mainly World War II veterans!) started training in 1955, and by the end of the year the first North American F-86F Sabre jet fighter squadron was operational. The JASDF strength by 1958 was almost three hundred Sabres, thirty-five Curtiss C-46 Commando transports, and some three hundred trainers. Many of the Sabres and T-33As were Japanese-built. In 1960, Japanese-built Lockheed F-104J Starfighters started to enter service, while Japanese-designed Fuji T-1 jet trainers had been delivered before this. McDonnell Douglas F-4EJ Phantom IIs first entered service in 1970.

Currently the JASDF has some 40,000 men, and operates seven squadrons with a total of 175 Lockheed F-104J Starfighter interceptors; eight squadrons of a total of two hundred North American F-86F Sabre and four squadrons with one hundred McDonnell Douglas F-4EJ Phantom II fighter-bombers; one squadron of twenty North American RF-86F Sabre reconnaissance-fighters; fifty transport aircraft including Curtiss C-46 Commando and NAMC YS-11 types; twenty Sikorsky H-19 and S-62, and Vertol 107 helicopters; Beech T-11 Kansan and T-34

Mentor, Fuji T-1, and Lockheed T-33A, and some Lockheed TF-104G Starfighter trainers. Three Nike-Ajax surface-to-air missile battalions are operated. In 1972 deliveries of the NAMC C-1 tactical jet troop-transport will begin – at present the orders are for about forty aircraft, while during the mid-1970s, one hundred Mitsubishi XT-2 supersonic trainers will also enter service.

Japanese Maritime Self-Defence Force

The Japanese Maritime Self-Defence Force came officially into existence in 1954, when it was formed primarily as an anti-submarine force. Four Bell TH-13 helicopters and some North American SNJ-6 trainers formed the initial equipment. By 1955 twenty Grumman TBM-3W-2 and TBM-3S Avengers, seventeen Lockheed PV-2 Harpoon, ten Convair PBY-6A and four Grumman Goose amphibians were being operated, with Sikorsky S-51 helicopters. In 1956, Lockheed P2V-7 Neptunes started replacing the Harpoons, and, the following year, the Avengers were replaced by sixty Grumman S-2A Tracker anti-submarine aircraft. Forty-two Neptunes were built in Japan.

Currently the JMSDF operates forty-six Kawasaki P-2J maritime-reconnaissance aircraft, which are turboprop developments of the Neptunes they are to replace by 1974: Three squadrons of Shin Meiwa PS-1 turboprop anti-submarine flying-boats will eventually be formed from the thirty-six currently on order. Four squadrons operate a total of fifty-six Grumman S-2A Trackers, one squadron operates fourteen Sikorsky SH-34J and three squadrons a total of fifty-six Mitsubishi-Sikorsky SH-3A Sea King helicopters for anti-submarine warfare. Eight Mitsubishi-Sikorsky S-62A and fourteen Kawasaki-Vertol 107 helicopters are also operated, along with four Douglas R4D-6 Dakotas, six NAMC YS-11s, thirty Beech 65 Queen Airs, five Fuji-Beech T-34 Mentor and twenty-seven Fuji KM-2s, and five Grumman HU-16 Albatross amphibians.

Japanese Ground Self-Defence Force

The United States Army trained a number of pilots in 1952 for the National Police Reserve, using Piper L-21 and Stinson L-5 Sentinels.

These pilots and aircraft formed the air element of the Japanese Ground Self-Defence Force when it came into existence in 1954.

Currently the JGSDF is just completing an expansion programme taking its helicopter force from 290 machines to four hundred. This force includes forty-two Kawasaki-Vertol KV107, seventy-nine Fuji-Bell UH-1B Iroquois, eighty-two Kawasaki-Bell H-13 Sioux, sixty Kawasaki-Hughes OH-6A, twelve Sikorsky H-19 Chickasaw and fourteen Mitsubishi-Sikorsky S-62 helicopters, with the possibility of further big orders for the OH-6A; nine Mitsubishi MU-2C light transports, and some Cessna O-1 Bird Dog AOP aircraft.

JORDAN

Royal Jordanian Air Force

Originally the Royal Jordanian Air Force was financed entirely by the United Kingdom as the Arab League Air Force, formed in 1949 after the end of the Arab-Israeli War with two de Havilland Rapides for light transport duties. RAF personnel were seconded to key posts, and two de Havilland Tiger Moth and four Percival Proctor trainers were added to the force by the end of the first year. Training of Arab personnel began in 1950, and a selection of Auster aircraft were obtained for AOP duties and to expand the training flight. Three Turkish-built M.K.E.K.4 Ugur trainers operated briefly before replacement by de Havilland Canada Chipmunks. During the next couple of years, four de Havilland Doves, a Vickers Viking and a Handley Page Marathon communications and VIP aircraft entered service, and were joined in 1956 by a Vickers Varsity. The first jet equipment, two de Havilland Vampire T.55 trainers and nine FB.9 fighter-bombers, were a gift of the British Government in 1955, and supplemented by ex-Egyptian FB.52s in 1956. Three ex-RAF North American T-6 Harvard trainers were also obtained in 1955.

The present title was assumed in 1956, and the RJAF became a separate service. After a brief flirtation with Egypt, Jordan, backed by Saudi Arabia, returned to her association with the United

Kingdom. Twelve Hawker Hunter F.6 fighters, four Sud Alouette III and a Westland Scout helicopters, with additional Doves and a Beech Twin Bonanza, arrived during the next few years.

During the June 1967 Arab-Israeli War, Jordan lost almost all of her aircraft, along with that part of her territory on the west bank of the River Jordan. Re-equipment was made with British, American and Pakistani aid (four North American F-86F Sabres on loan).

Currently the 2,000-strong RJAF operates thirty-six Lockheed F-104A Starfighter fighter aircraft in two squadrons; a squadron of eighteen Hawker Hunter FGA.9 fighter-bombers; four Sud Alouette III and four Westland S55 Whirlwind helicopters; four Douglas C-47 Dakota, two Hawker Siddeley Dove and two Heron transports; and a number of de Havilland Canada Chipmunk trainers. Short Tigercat surface-to-air missiles are used for airfield defence. Pilot training takes place in the United States.

KENYA

Kenya Air Force

Kenya attained independence from the United Kingdom in 1963, and since then has received British aid and assistance in building up national defence forces. Currently, seven de Havilland Canada DHC-2 Beaver and four DHC-4 Caribou transports are operated, along with an Aero Commander 500 VIP aircraft and six de Havilland Canada Chipmunk trainers. Five Scottish Aviation Bulldog trainers are on order for 1972 delivery, while six BAC 167 Strikemaster armed-trainers entered service in late 1971.

REPUBLIC OF KOREA (SOUTH)

Republic of Korea Air Force

Korea was occupied from 1945 onwards, and the country was divided into a Communist zone and an American zone: North Korea and South Korea respectively. Before this, and since 1910,

Korea had been a Japanese protectorate. The Republic of Korea was formed in 1948, and the following year the Republic of Korea Air Force was formed using Koreans who had flown with the Japanese forces during World War II. Initial equipment consisted of Piper L-4 liaison and AOP aircraft, and North American T-6 Harvard trainers.

In June 1950, North Korean forces invaded South Korea, marking the beginning of the Korean War in which United Nations forces, mainly drawn from the United States, the United Kingdom and Australia, opposed Communist forces. Rapid expansion of the ROKAF followed; North American F-51D Mustang fighters were placed in service almost immediately. After the war ended in 1953, American military aid continued at a high level. In 1954, Aero Commander 500 light transports were obtained, and in 1956 North American F-86F Sabre jet fighters became operational, and Lockheed T-33A jet trainers arrived in 1957, followed by F-86D Sabres. In 1965, Northrop F-5A/B Freedom Fighters started to arrive, and in 1970, McDonnell Douglas F-4D Phantom IIs entered service.

Currently the ROKAF is 23,000 men strong, and operates eighteen McDonnell Douglas F-4D Phamton II, one hundred North American F-86F Sabre and forty Northrop F-5A fighter bombers; a further twenty F-86D Sabres operate in the interceptor role, while ten RF-86F Sabres operate in a reconnaissance-fighter squadron. Douglas C-47 Dakota, Curtiss C-46 Commando and Aero Commander 500 transports are operated; with six Sikorsky H-19 Chickasaw and five Bell UH-1D Iroquois helicopters; Lockheed T-33A, Northrop F-5B, North American T-6 Harvard and T-28A trainers; and Cessna O-1 Bird Dog AOP aircraft. Hawk surface-to-air missiles are deployed.

DEMOCRATIC PEOPLE'S REPUBLIC OF KOREA (NORTH)

Korean People's Army Air Force

The Korean People's Army Air Force has its origins in the North Korean Aviation Society which was formed after Soviet occupation of North Korea in 1945, and in 1946 this force became the Korean People's Army Aviation Division, with yet a further change of name in 1948 to the Korean People's Armed Forces Air Corps. It was at this time that Russian assistance reached significant proportions, with Yakovlev Yak-9 fighters, Ilyushin Il-10 ground-attack aircraft, and Yak-18 trainers, delivered between 1948 and June 1950 when North Korean forces invaded the South. Towards the end of 1950, Mikoyan MiG-15 jet fighters started to enter service, and this type bore the brunt of the aerial action during the war.

After the war ended in 1953, the Democratic People's Republic of Korea was formed and the title of the Korean People's Army Air Force was adopted for the air arm. On its formation the KPAAF was operating Mikoyan MiG-15 jet fighters, Yakovlev Yak-9 and Lavochkin La-9 fighters, Tupolev Tu-2 bombers and Ilyushin Il-10 ground-attack aircraft, with Yak-11 and Yak-18 trainers. Ilyushin Il-28 jet bombers followed, and in 1957, MiG-17s were delivered, including some of the Chinese-built Shenyang F-4, with Antonov An-2 and Lisunov Li-2 (C-47) transports. MiG-21 interceptors were delivered in 1966.

Currently the KPAAF, which has some 30,000 men, operates ninety Mikoyan MiG-21 interceptors; twenty MiG-19, 340 MiG-17 and sixty MiG-15 fighter-bombers (some of these are undoubtedly Shenyang versions); seventy Ilyushin Il-28 bombers; thirty Antonov An-2, Lisunov Li-2, Ilyushin Il-12 and Il-14 transports; twenty Mil Mi-4 and some Mi-1 helicopters; with Yakovlev Yak-11 and Yak-18, MiG-15UTI and Il-28U trainers.

KUWAIT

Kuwait Air Force

One of the many small air forces formed with British assistance, the Kuwait Air Force first came into being in 1960 as an extension of the Security Department; at its formation it operated de Havilland Doves and a Heron, and five Austers. It is now a highly sophisticated air force with twelve BAC Lightning interceptors, and two conversion trainers, six BAC 167 Strikemaster armed-trainers, six Hawker Hunter ground-attack aircraft, six Agusta-Bell 204B and two Westland S-55 Whirlwind helicopters, two de Havilland Canada DHC-4 Caribou transports, and six BAC Jet Provost trainers.

LAOS

Royal Lao Air Force

Laos became an independent member of the French Union in July 1949, and was in fact an amalgamation of several French colonies. Four years later the country was invaded by communist Viet Minh forces and, although these withdrew after a period, the country has since been subjected to attempts by communist Pathet-Lao forces to overthrow the Government.

It was in 1954 that the Laotian Army Aviation Service was formed, with American assistance, for counter-insurgency duties. Initial equipment consisted of twenty Cessna O-1 Bird Dog AOP aircraft; ten Douglas C-47 Dakota, three de Havilland Canada DHC-2 Beaver, and four Aero Commander 520 transports. Sikorsky S-55 helicopters and North American T-6G Texan trainers were delivered later, with North American T-28D armed-trainers for counter-insurgency duties. In 1960 the present title was adopted.

Currently the RLAF operates some sixty North American T-28D armed-trainers; ten Douglas C-47 Dakota, three de Havilland Canada DHC-2 Beaver and some Aero Commander 520 transports; Sud Alouette II and Alouette III, Sikorsky UH-19 and UH-34 helicopters; with North American T-6G Texan trainers.

LEBANON

Lebanese Air Force *Force Aérienne Libanaise*

Prior to 1918 the Lebanon was under Turkish rule, and from 1918 to 1943 the country was administered by France under a

League of Nations Mandate. Independence was granted in 1943, and in 1949 the Force Aérienne Libanaise was formed, with British assistance, as an internal security force. The initial equipment consisted of two Percival Prentice trainers, but these were soon joined by three Savoia-Marchetti S.M.79 tri-motor transports, a de Havilland Dove light transport and some de Havilland Canada Chipmunk trainers in 1950. Former French Armée de l'Air bases were used, and, by 1955, de Havilland Vampire FB.52 jet fighter-bombers and T.55 jet trainers, North American T-6 Harvard and additional Chipmunk trainers, and an Aermacchi MB.308 communications aircraft were in service. A few ex-Iraqi de Havilland Tiger Moths were also supplied for AOP duties.

During the late 1950s five Hawker Hunter F.6 fighters and five FB.9 fighter-bombers entered service, and have now been supplemented by twelve Dassault Mirage IIIC fighter-bombers.

Currently the FAL has 1,000 men, and operates one squadron with twelve Dassault Mirage IIIC fighter-bombers; one squadron with five Hawker Hunter F.6 fighters and five Hawker Hunter FGA.9 fighter-bombers; a few transport aircraft, including a de Havilland Dove; three Sud Alouette II and seven Alouette III helicopters in one squadron; four Potez Super Magister, two Dassault Mirage IIIBC, and eleven de Havilland Canada Chipmunk trainers. A French Crotale air defence system with low-level surface-to-air missiles has just been installed.

LIBYA

Libyan Air Force

The Libyan Air Force was formed in 1959 with two trainers supplied by Egypt. Additional equipment consisted of two Auster AOP 6s provided by the United Kingdom, and two Lockheed T-33A jet trainers and a Douglas C-47 Dakota transport provided by the United States in 1963. These aircraft were part payment for the use of Libyan bases. An arms deal with the United Kingdom which included surface-to-air missiles was scrapped after a change of government in Libya in 1970. However, prior to this, seven Northrop F-5A fighter-bombers were delivered. Defence procurement is now likely to be from French and Soviet bloc sources.

Currently there are nine Northrop F-5A and sixty Dassault Mirage IIIE fighter-bombers now entering service, and fifty Mirage 5 ground-attack aircraft for 1972 delivery; six Lockheed C-130E Hercules, nine Douglas C-47 Dakota and a Dassault Falcon transports; ten Sud Alouette II and Super Frelon helicopters; and MiG-21UTI and Lockheed T-33A jet trainers.

MALAGASY REPUBLIC

Malagasy Air Force *Armée de l'Air Malgache*

The former French colony of Madagascar became independent in 1960, and remained a member of the French Community after that date. It is currently operating a small air force, with 200 men, which has three Douglas C-47 Dakotas, six Max Holste 1521M Broussards, and two Dassault M.D.315 Flamants, with a Bell 47G Sioux and a Sud Alouette III helicopters.

MALAYSIA

Royal Malaysian Air Force

Malaysia came into existence in September 1963, being the former British colonies of the Federation of Malaya, the State of Singapore, Sabah (or North Borneo) and Sarawak, although Singapore withdrew in 1965.

Although the Royal Malaysian Air Force itself dates from 1958, Malaysian military aviation dates from 1936 with the formation of the Straits Settlements Volunteer Air Force, which was equipped with Hawker Audax aircraft. In 1940 this became the Malayan Volunteer Air Force, with civil aircraft, including de Havilland Tiger and Leopard Moths and Rapides, requisitioned on the outbreak of World War II in the previous year. Although this force escaped from the Japanese invasion of

Malaya, the fall of Sumatra which followed saw it defeated by overwhelming odds.

During the immediate post-war period, military aviation in Malaya was left to the British forces, but in 1950 a Malayan Auxiliary Air Force was formed as a training organization equipped with de Havilland Tiger Moths. North American T-6 Harvard trainers soon followed, and then Supermarine Spitfire F.21 fighters were operated for a short spell. De Havilland Canada Chipmunk trainers appeared in 1957.

Independence for Malaya (but not for the other territories) in 1958 led to the formation of the Royal Malayan Air Force equipped with Scottish Aviation Twin Pioneer transports and de Havilland Canada Chipmunk trainers. The title Royal Malaysian Air Force was adopted after the formation of the Federation of Malaysia, and a number of new aircraft arrived at this time, including Handley Page Herald and de Havilland Canada DHC-4 Caribou transports, and de Havilland Dove, Devon (military Doves) and Heron light transports. These were joined by Canadair CL-41 Tutors, which were known locally as the Tebuan, or Wasp. RMAF aircraft and personnel fought alongside the RAF and Fleet Air Arm against Indonesia during the period immediately following the formation of the Federation, when the Indonesians tried to destroy it by a policy of confrontation which involved infiltrating guerillas into the Federation. The RAF, RAAF, and RNZAF have bases in Malaysia, and the Federation co-operates with Singapore on defence matters. Commonwealth CA-27 Avon-Sabres were delivered in 1969.

Currently, the RMAF has 4,500 men and operates ten Commonwealth CA-27 Avon-Sabre fighter-bombers with a squadron of Dassault Mirage Vs on order; twenty Canadair CL-41G Tebuan armed-trainers; ten Handley Page Herald 401, twelve de Havilland Canada DHC-4 Caribou, fourteen Scottish Aviation Pioneers and Twin Pioneers, five de Havilland Doves and two Herons for transport and communications duties; twenty Sud Alouette III and ten Sikorsky S-61A helicopters; and eighteen BAC Provost T-51 and fifteen Scottish Aviation Bulldog trainers.

MALI REPUBLIC

Mali Air Force *Force Aérienne du Mali*

Mali became independent of France in 1960, but remained a member of the French Community for four years before the link was severed. Initial equipment included the French standard package of a Douglas C-47 Dakota and two Max Holste 1521M Broussards. The United States supplied two more C-47s, while the Soviet Union is supposed to have provided six obsolete Mikoyan MiG-15 jet fighters.

MAURITANIA

Mauritanian Islamic Air Force
Force Aérienne de la Republique Islamique de Mauritanie

A former French colony, Mauritania became independent in 1960; but although certain links with France were maintained, Mauritania stayed outside of the French Community. The standard package of a Douglas C-47 transport and two Holste 1521M Broussards is operated.

MEXICO

Mexican Air Force *Fuerza Aérea Mexicana*

Mexico was one of the first Latin American countries to take an interest in military aviation: as early as 1910 various revolutionary and counter-revoluntionary groups employed aircraft with mercenary pilots for AOP duties. In 1915 the Government founded an aircraft industry to overcome the shortage of outside supplies caused by World War I.

A more formal approach came in 1924 with the formation of the Fuerza Aérea Mexicana to operate American-built D.H.4B bombers and Douglas O-2 AOP aircraft, with some Bristol general-purpose biplanes, on police duties. In 1930, Avro 504K trainers were added and later some Azcarate-E trainers, a national design, were also introduced. Although ten Vought O2U Corsair biplanes were obtained during the early 1930s, most of the decade saw little further progress until a few Grumman G-23 fighters, Waco D-6 general-purpose aircraft, and Consolidated Fleet 21 and Ryan S-T trainers were obtained during the immediate pre-war period.

In 1942, Mexico joined the Allies, putting bases at their disposal. In return, the United States supplied Douglas A-24 Dauntless anti-submarine aircraft, Vought-Sikorsky OS2U AOP and North American NA-16 training aircraft. Mexico produced the nationally-designed Tezuitlan trainer during the war years. In 1945 a FAM squadron equipped with Republic F-47D Thunderbolt fighter-bombers was sent to fight with Allied forces against Japan, but surrender came before its arrival. After the war, Douglas C-47 Dakota, Lockheed Lodestar and Beech C-45 transports were received, along with trainers of Fleet, Fairchild and Vultee manufacture.

The late 1950s saw the arrival of de Havilland Vampire F.3 fighters and T.55 trainers, followed in 1961 by Lockheed T-33A armed-trainers. Mexico is a member of the Organization of American States.

Currently the FAM, which has 6,000 men, operates fifteen de Havilland Vampire F.3 and fifteen Lockheed T-33A aircraft as fighter-bombers; while forty-five North American T-6G Texan and thirty T-28A Trojan armed-trainers are also available for COIN duties. There are six Douglas C-47 Dakota, five C-54 and two C-118A, and some Beech C-45 transports; eighteen Bell 47G Sioux, a Hiller UH-12E and eight Sud Alouette II helicopters; while for training there are (in addition to the T-6s and T-28s already mentioned) fifteen Beech T-11 Kansan and ten T-34 Mentor aircraft.

Naval Aviation   *Armada da Mexico*

Mexican Naval Aviation currently operates fourteen aircraft on search and rescue duties, including five Consolidated PBY-5A Catalina amphibians, five Sud Alouette II and four Bell 47J Sioux helicopters.

MONGOLIA

Air Force of the Mongolian People's Republic

Mongolian military aviation dates from 1926, when four Russian-built aircraft were supplied to the Army. A flying school was formed in 1927, but there is little factual history about the early years of the air arm, although about a hundred aircraft of Russian design were in service by 1933 and Russian aid was forthcoming to counter the growing Japanese presence in the area during the 1930s. A peak strength of 450 aircraft was reached in 1938, and at this time there were Polikarpov I-15 and I-16 fighters, R-5 AOP aircraft, and Tupolev TB-3 bombers. The only aerial combat with Japanese forces took place briefly in 1939, while during World War II a Mongolian Lavochkin La-5 fighter squadron fought against the Germans.

Post-war aircraft included Polikarpov Po-2s and Antonov An-2s, followed during the 1950s by Ilyushin Il-14 and Lisunov Li-2 (C-47) transports, and Yakovlev Yak-11 and Yak-18 trainers; and Mikoyan MiG-15 jet fighters eventually entered service.

Currently strength is seven hundred men, and ten Mikoyan MiG-15 fighter-bombers are operated; along with Antonov An-2 and An-12, Ilyushin Il-12 and Il-14 transports; Mil Mi-4 helicopters; and Yakovlev Yak-11 and Yak-18 and UT-2 trainers.

MOROCCO

Royal Moroccan Air Force *Aviation Royale Chérifienne*

Morocco became independent of both France and Spain in 1956, and, in that same year, Aviation Royale Chérifienne was formed. At first personnel were trained in France and Spain, and Armeé de l'Air personnel were seconded to the ARC which had as initial equipment six Morane-Saulnier M.S.500 Criquet AOP aircraft. A de Havilland Heron, two Beech Twin Bonanza and three Max Holste 1521M Broussard transports were delivered in 1957, with a Bell 47 Sioux helicopter.

In 1961 the USSR started giving aid to Morocco, including Mikoyan MiG-17 fighter-bombers, Ilyushin Il-28 jet bombers and MiG-15UTI trainers. A fair amount of American equipment has also been received, however, including Northrop F-5A fighter-bombers in 1966, and also Fairchild C-119G Packet and Douglas C-47 Dakota transports. Recently the USSR has turned its attentions to Algeria, and the MiG-17s are in reserve, possibly through a lack of spares.

Currently the ARC has some 4,000 men, and operates one squadron with ten Northrop F-5A fighter-bombers; twenty-four Potez Magister (delivered 1970) armed jet trainers; twenty-five North American T-28 Trojan armed-trainers; ten Douglas C-47 Dakota and eleven Fairchild C-119G Packet transports, with some Beech Twin Bonanza and Max Holste 1521M Broussard aircraft for light transport and communications duties; thirteen Agusta-Bell 205, four Bell 47 Sioux, some Sikorsky H-34 and Kaman HH-43 Huskie helicopters; with forty-five North American T-6G Texan and eight Potez Magister trainers.

MUSCAT AND OMAN

Sultan of Oman's Air Force

The Sultan of Oman's Air Force was formed in 1959 with British assistance, for counter-insurgency and police duties. Largely manned by RAF personnel on secondment and by some ex-RAF personnel, the SOAF operates twelve BAC 167 Strikemasters and five Provost T.52 armed-trainers, four de Havilland Canada DHC-2 Beaver and six Short Skyvan 3M transports.

NEPAL

Royal Flight and Army Aviation

Although the small state of Nepal, home of the Gurkha soldiers, does not have a formal air force, a Sud Alouette III helicopter was operated as air transport for the royal family, and in 1970

1 Argentinian Air Force BAC Canberra
2 Argentinian Navy de Havilland Canada DHC-6 Twin Otter

3 Royal Australian Air Force Dassault Mirage III-Os

4 Royal Australian Navy Westland Wessex

5 Royal Australian Navy Grumman S-2E Tracker and Douglas A-4G Skyhawks

6 Austrian Air Force SAAB 105E

7 Austrian Air Force Sud Alouette III

8 Austrian Air Force Sikorsky S-65S

9 Belgian Air Force Dassault Mirage 5

10 Belgian Army Dornier Do 27 and Sud Alouette II helicopter

11 Canadian Armed Forces Grumman HU-16 Albatross

12 Canadian Armed Forces de Havilland Canada DHC-5 Buffalo

13 Colombian Air Force Douglas C-47 Dakota

14 Colombian Air Force North American F-86F Sabre
15 French Air Force Dassault Mirage F.1 and G

16 French Air Force Dassault Mirage IV

17 French Army Max Holste 1521M Broussard

18 French Army Sud SA.330 Puma

19 USAF, RAF and Luftwaffe McDonnell Douglas F-4 Phantom IIs

20 Dassault-Breguet-Dornier Alpha Jet
21 West German Army Dornier Do 27

22 West German Air Force Lockheed F-104G Starfighters

23 Indian Air Force Hawker Siddeley Gnat

24 Indian Air Force BAC Canberra

25 Italian Air Force Fiat G.91s

26 Italian Air Force Agusta-Bell 204B Iroquois

27 Japanese Air Self-Defence Force NAMC XC-1A

28 Japanese Air Self-Defence Force Lockheed F-104J Starfighter

29 Kuwait Air Force BAC 167 Strikemaster

30 Royal Malaysian Air Force Sikorsky S-61A

31 Royal Netherlands Navy Breguet Br.1150 Atlantic

32 Royal Netherlands Air Force Fokker F-27M Troopships

33 Polish Air Force Sukhoi Su-7Bs

34 Polish Air Force Mikoyan MiG-21PFs
35 Polish Air Force Mikoyan MiG-17s

36 South African Air Force Aermacchi MB.326K
37 South African Air Force Hawker Siddeley Shackleton MR.3

38 South African Air Force Lockheed C-130B Hercules

39 South African Air Force Dassault Mirage III-BZ

40 Swiss Air Force Dassault Mirage III-RS

41 Russian Tupolev Tu-20 and USAF General Dynamics F-102 Delta Dagger

43 Kuwait and Saudi Arabian Air Force BAC Lightnings

44 Panavia 200 Panther

Opposite
42 Russian Navy Kamov Ka-25

45 Royal Air Force BAC-Breguet Jaguar

46 Royal Air Force Hawker Siddeley Harrier

47 Royal Air Force Hawker Siddeley Hunter

48 Royal Air Force Hawker Siddeley Nimrod

49 Royal Navy Westland SH-3D Sea King

Opposite
50 Royal Navy Westland Wasp
51 British Army Westland-Bell 47
52 British Army Westland Scout
53 British Army Westland WG.13

54 McDonnell Douglas F-15

55 United States Air Force General Dynamics F-111A

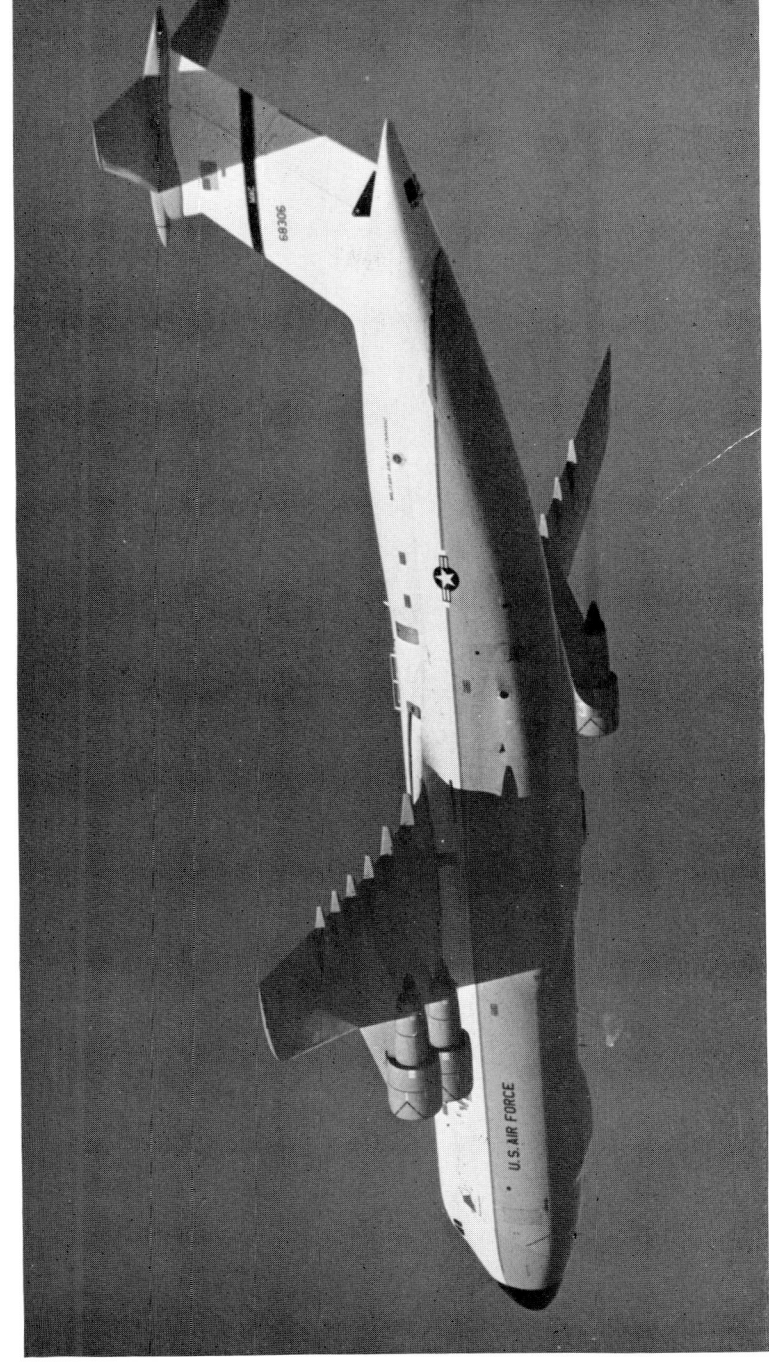

56 United States Air Force Lockheed C-5A Galaxy

57 United States Air Force Northrop F-5A Freedom Fighter and T-38 Talon

58 United States Air Force Cessna A-37

59 Douglas A-1E Skyraiders, Lockheed KC-130H Hercules and Sikorsky CH-3B

60 United States Air Force Cessna O-2A, Bell UH-1D Iroquois and Sikorsky S-58

61 United States Air Force Sikorsky CH-3C

62 United States Navy Ling-Temco-Vought A-7E Corsair II

63 United States Navy McDonnell Douglas F-4B Phantom II

64 United States Navy Grumman E-2 Hawkeye

65 United States Navy Lockheed P-3 Orion

66 United States Navy Kaman UH-2 Seasprite

67 United States Marine Corps Bell AH-1J Sea Cobra

68 United States Marine Corps Hawker Siddeley AV-8 Harrier

69 United States Marine Corps McDonnell Douglas RF-4B Phantom II

70 United States Army Boeing-Vertol CH-47 Chinook

this was supplemented by a Short Skyvan 3M VIP transport. In 1971 the Royal Nepalese Army received its first aircraft, also a Short Skyvan 3M, but in the utility version.

NETHERLANDS

Royal Netherlands Air Force *Koninklijke Luchtmacht*

Although an artillery observation balloon was operated for a brief period in 1886, Dutch military aviation did not really start until 1911, when aircraft and balloons were obtained for manoeuvres. Two years later the Royal Netherlands Army formed an Aviation Division, which operated two van Meel and three Farman F-22 aircraft. The following year, 1914, the Royal Netherlands Indies Army obtained two Martin TT seaplanes, a further twelve being placed in service in 1917. Meanwhile the Royal Netherlands Army had obtained more Farman F-22s, so that by 1915 there were twenty of this type in service. Twenty Nieuport 17C-1 Bébé and ten Fokker D.III fighters were obtained in 1917, with additional Nieuports and some Caudron G.III reconnaissance aircraft in 1918. After World War I ended (the Netherlands had remained neutral), forty ex-German Rumpler C.Vs and thirty-six Trumpenburg Spijke trainers were bought.

Ambitious post-war re-equipment plans were pruned drastically, with new equipment limited to sixteen Thalin K fighters, twenty Fokker D.VII fighters and fifty-six C.I. reconnaissance aircraft. The 1920s were in fact fairly difficult for the Aviation Division: during the period only a further fifteen Fokker D.XVI and ten D.XVII fighters, thirty-two C.IV, one hundred C.VI, four C.VIII and five C.IX reconnaissance aircraft, three F.VIIa trimotor transports, and thirty S.IV trainers were put into service – and lasting for much of the 1930s as well! The Royal Netherlands Indies Army fared slightly better as far as modern equipment, if not actual numbers, was concerned. After the war, twenty-five each of D.H.9 bombers and Avro 504K trainers were received, while during the 1920s ten Vickers Viking amphibians, six Fokker D.VII and four D.C.I fighters, and twenty C.IV reconnaissance aircraft were placed in service. In 1930 nine Curtiss P-6E Hawk fighters were delivered, followed during the late 1930s by more than a hundred Martin 139-W bombers.

During the inter-war years a policy of neutrality and a belief in the inevitability of defeat contributed to the situation already mentioned, that of obsolete aircraft, and very few of them!

A last-minute attempt to rectify the situation was made in 1938. The Aviation Division became the Army Air Service in a reorganization. Orders were placed for Dutch and American aircraft, but at the time of Germany's invasion of the Netherlands on 10 May 1940 the Army Air Service could only muster thirty Fokker D.XXI, some D.XVII and twenty G.1 fighters; sixteen T.V bombers; eleven Douglas DB-8A attack aircraft; forty Fokker C.V and C.X reconnaissance aircraft; and a few Koolhaven trainers in service as AOP aircraft. This pitiful force fought valiantly for five days before defeat, after which the flying school flew to England.

The Royal Netherlands Indies Army fared rather better again, receiving many aircraft which were too late to be of any use for the parent service, including twenty-four Curtiss Hawk 75-A, twenty-four Curtiss-Wright CW-22 Falcon, and seventy-two Brewster 339 and 439 fighters; a few Douglas DB-7 bombers; thirty-six Consolidated PBY-5 Catalina and forty-eight Sikorsky S-43 amphibians; and twenty Lockheed Lodestar transports. After the Netherlands declared war on Japan in December 1941, these aircraft and the Martin 139-W bombers were operated on missions in support of British forces in Borneo and Singapore, but by the following March the remnants were in Australia and Ceylon, continuing the fight after the Japanese had invaded the Netherlands Indies.

In Europe, Netherlands squadrons operated alongside the RAF and Royal Navy. These were to form the nucleus of post-war Dutch military aviation after the liberation of the Netherlands in October 1944, and they operated Supermarine Spitfire fighters; North American B-25 Mitchell bombers; Douglas C-47 Dakota transports and Auster AOP 3s. In 1947, North American T-6 Harvard, de Havilland Dominie and Tiger Moth, Percival Proctor, Avro Anson and Airspeed Oxford trainers were delivered – some 350 in all – along with Lockheed 12 and 14 transports. The Royal Netherlands Indies Army was dealing with Indonesian rebels using North American F-51D Mustang and Curtiss P-40N Warhawk fighters, Mitchells and Dakotas. In 1948 the Army Air Service received the first of two hundred Gloster Meteor F.4 and F.8 jet fighters – many of these aircraft were licence-built in the Netherlands. Forty Fokker S.11 trainers were also received in the same year.

The Netherlands became a member of the North Atlantic Treaty Organization in 1952.

In 1953 the Army Aviation Service became the Royal Netherlands Air Force, a separate service, and started to receive the first of two hundred Republic F-84E Thunderjet fighter-bombers, which were later replaced by F-84F Thunderstreaks, Hawker Hunters and North American F-86F Sabres. Also during the early 1950s, Piper L-18 Super Cubs, de Havilland Canada DHC-2 Beavers, Hiller H-23 helicopters and twenty Fokker S.14 jet trainers were delivered. Lockheed T-33A trainers arrived in due course.

The first of 120 licence-built Lockheed F-104G Starfighter fighter-bombers entered service in 1963, and 105 Thunderstreak-replacement Northrop F-5A/Bs started to arrive later in the decade – these being Canadian-built versions known as NF-5A/Bs. Nike-Ajax and Nike-Hercules surface-to-air missiles also were first deployed during this period.

Currently the RNAF has some 21,000 men, and operates four Lockheed F-104G squadrons with twenty aircraft each; two squadrons in the interceptor role, and two as fighter-bombers; one Lockheed RF-105G photo-reconnaissance squadron; three Northrop NF-5A fighter-bomber squadrons with a total of seventy-five aircraft; twelve Fokker F-27 Troopships in one transport squadron; one squadron of de Havilland Canada DHC-2 Beaver transports, which is under Army Command, as are most of the RNAF's seventy-seven Sud Alouette III helicopters and sixty Piper L-21 Super Cubs. Training is on thirty Northrop NF-5Bs which replaced the T-33As, twenty Hawker Hunter T.7s, ten Lockheed TF-104Gs and forty Fokker S-11 Instructors. Eleven squadrons of Nike-Ajax and Nike-Hercules surface-to-air missiles have been replaced with Hawk missiles. Squadron strengths may be cut during the early 1970s as part of a defence cut programme.

Royal Netherlands Naval Air Service *Marine Luchtvaartdienst*

The Marine Luchtvaartdienst was formed in 1917 with six Martin seaplanes and three Farman F-22 aircraft. At first most attention was paid to the Netherlands Indies, whose many islands were an obvious setting in which naval aviation could prove its worth. The MLD operated some forty Hansa-Brandenburg W-12 seaplanes in the Netherlands Indies during the 1920s, and these were supplemented by Dornier Wal flying-boats, which remained in service until 1942 when Japan invaded. Before World War II started, some Dornier Do 24K flying-boats entered service in the Indies, while in the Netherlands units were equipped with Fokker T.VIIIW seaplanes, which escaped to France after the fall of Holland in 1940 and from there moved to the United Kingdom.

The Netherlands declared war on Japan in 1941, but when Japan invaded the following year, the Dutch forces, spread over a wide area, were defeated quickly, although many did manage to escape to Ceylon and Australia. MLD personnel in the United Kingdom flew Fairey Swordfish torpedo-bombers, and, after the liberation of the Netherlands in 1944, Fairey Barracuda torpedo-bombers were operated before thirty Fairey Fireflies replaced these. In 1946 a Royal Navy escort carrier HMS *Nairana* entered Dutch service on loan until the Royal Netherlands Navy's own light fleet carrier, *Karel Doorman*, formerly HMS *Venerable*, entered service in 1948. Initially, later versions of the Firefly were operated from the *Karel Doorman*, with Hawker Sea Furies, followed by Grumman TBM-3W and TBM-3S Avengers for anti-submarine duties and Hawker Sea Hawk jet fighter-bombers; while eventually three squadrons of Grumman S-2 Trackers were placed in service.

Meanwhile, shore-based aircraft included a squadron of Convair PBY-5 Catalina amphibians and, in 1951, Lockheed PV-2 Harpoon maritime-reconnaissance aircraft, which were soon replaced by P2V-5 Neptunes in 1953.

The *Karel Doorman* was sold to Argentina in 1969, leaving eight guided-missile frigates of the Von Speijk class operating Westland Wasp helicopters. The fifteen Lockheed Neptunes have been replaced in 1970 by nine Breguet Br.1150 Atlantic maritime-reconnaissance aircraft.

Currently the MLD operates one squadron with nine Breguet Br.1150 Atlantic maritime-reconnaissance aircraft; three squadrons with a total of thirty-six Grumman S-2A Tracker anti-submarine aircraft, now all shore-based; twelve Westland Wasp helicopters, operated from warships; eight Sikorsky SH-34J and six Bell UH-1 Iroquois helicopters.

NEW ZEALAND

Royal New Zealand Air Force

Although New Zealand did not have any air arm of her own until 1923, when the New Zealand Permanent Air Force was formed, New Zealanders played an active part in World War I as members of the Royal Flying Corps, Royal Naval Air Service and, ultimately, the Royal Air Force. The moving force behind the decision to form the NZPAF was the 1920 offer by the British Government to the colonies of surplus RAF aircraft. New Zealand's share of the Imperial Gift was to be one hundred aircraft, but delays in deciding what to do with these meant that there was little choice left by the time a decision was taken, and less than half of the number offered were in fact taken up. A few aircraft were leased to civilian operators in New Zealand, while the NZPAF received ten Bristol F.2B fighters, ten D.H.4 and nine D.H.9 bombers, and four Avro 504K trainers. The NZPAF was a part of the Army, while a reserve force, the New Zealand Air Force, used NZPAF aircraft for training.

Aircraft entering service during the 1920s were few, due to the economic difficulties of the period, and included five Gloster Glebe and de Havilland Puss Moth and Moth trainers, supplemented by a few additional Avro 504K trainers. Some expansion occurred in the 1930s: a Saunders-Roe Cutty Sark flying-boat and ten Fairey IIIFs were followed in 1935 by twelve Hawker Vildebeest torpedo-bombers and some Avro 626 trainers. The title of Royal New Zealand Air Force came into use in 1934, but the force remained a part of the Army until 1937 when it became a separate service on the recommendation of an RAF officer seconded to New Zealand to advise on the future structure of the RNZAF.

Prior to the outbreak of World War II, a major expansion programme was put in hand, including the acquisition of new aircraft: ex-RAF Blackburn Baffins for the reserve squadrons, Airspeed Oxford trainers, Fairey Gordons, Hawker Harts and Vickers Vincents. An order for thirty Vickers Wellington bombers was due for delivery at the start of the war, but these were handed over to the RAF with crews as the nucleus of a New Zealand element. Basically the role of the RNZAF during the war was training of pilots and aircrew, mainly on North American T-6 Harvards. This did not prevent the RNZAF from fielding twenty-seven squadrons: twelve with fighters, initially Curtiss Kittyhawks and later Chance-Vought FG.1 Corsairs; six with bombers, Lockheed Hudsons followed by Venturas; two with flying-boats, initially with Short Singapores followed by Consolidated PBY-5 Catalina amphibians and Short Sunderland flying-boats; two Vickers Vincent general reconnaissance squadrons; a Douglas Dauntless dive-bomber squadron, and a Grumman Avenger squadron; with the remainder operating a variety of types including some Douglas C-47 Dakota and Lockheed Lodestar transports. For the most part the RNZAF fought in the Pacific, although some personnel flew with the RAF in Europe and North Africa. After the war, the RNZAF formed part of the British Commonwealth Occupation Force in Japan.

Post-war strength of the RNZAF was reduced to five regular and four reserve squadrons. During the early 1950s the reserve squadrons were equipped with North American F-51D Mustang fighters; while, of the regular squadrons, one operated de Havilland Vampire FB.9 jet fighter-bombers, another two operated de Havilland Mosquito bombers until these were replaced by one squadron of English Electric Canberra B(I) 12 jet bombers; a fourth operated Short Sunderland flying-boats for maritime-reconnaissance, until replaced by Lockheed P-3B Orions in 1967; and another squadron operated Douglas C-47 Dakota, Bristol 170M Freighters and Handley Page Hastings C.3 transports. The reserve squadrons were disbanded in 1957, at which time Harvards were being operated. A squadron of de Havilland Venom fighter-bombers was leased from the RAF for a few years.

Currently the RNZAF, which has some 4,500 men, operates one squadron of ten McDonnell Douglas A-4K Skyhawks; one squadron of twelve de Havilland Vampire FB.9s; one squadron of six BAC Canberra B(I) 12 bombers; one squadron with five Lockheed P-3B Orion maritime-reconnaissance aircraft; five Lockheed C-130H Hercules, fifteen de Havilland Devon and nine Bristol 170M Freighter transports; with fourteen Bell UH-1H Iroquois and thirteen Bell 47G Sioux helicopters. Training is on Douglas TA-4K Skyhawks, BAC 167 Strikemasters and Victa Airtourers. New Zealand is a member of the South-East Asia Treaty Organization, and a RNZAF squadron is based in Singapore.

Royal New Zealand Navy

The Royal New Zealand Navy operates two Westland Wasp helicopters, one of which is based aboard a general-purpose frigate at any one time.

NICARAGUA

National Guard Air Corps   *Fuerza Aérea, Guardia Nacional*

The Fuerza Aérea, Guardia Nacional was formed in 1938 with American help as a part of the National Guard. In 1948, Nicaragua became a member of the Organization of American States, and received aid from the United States in the form of Douglas C-47 Dakota transports. One squadron each of North American F-51D Mustang and Republic F-47D Thunderbolt fighter-bombers had been formed immediately after the end of World War II. In recent years, aircraft have been supplied by the United States for counter-insurgency duties.

Currently the Fuerza Aérea has some 1,500 men and operates one squadron with six Douglas B-26 Invader bombers; one squadron with six Lockheed T-33A armed-trainers; four Beech C-45, three Douglas C-47 Dakota and ten Cessna 180 transport and communications aircraft; one Hughes 269 and four OH-6A helicopters. North American T-6G Texan and T-28 Trojan, and Beech AT-11 Kansan trainers are also operated.

NIGER

Niger Air Force *Force Aérienne de Niger*

A former French colony, since 1960 Niger has maintained close links with France while being outside of the French Community. The air force consists of the standard 'package' of a Douglas C-47 Dakota transport and three Max Holste 1521M Broussards.

NIGERIA

Federal Nigerian Air Force

The Federation of Nigeria was formed in 1960 on being granted independence by the United Kingdom. The Federal Nigerian Air Force was founded in 1964 with Indian and German assistance. Initial equipment included ten Nord Noratlas transports, thirty Dornier Do 27 liaison aircraft, and twenty-six Piaggio P.149D trainers. During the late 1960s a state of civil war existed in Nigeria in which air power played little part, but it marked the start of Soviet military aid to the Federal Government, and a few aircraft were flown by Egyptian pilots.

Currently the FNAF operates ten Mikoyan MiG-17C and a handful of MiG-15 fighters; three Ilyushin Il-28 bombers; three Douglas C-47 Dakota transports; twenty Dornier Do 27 liaison aircraft; three Westland Whirlwind and seven Sud Alouette II helicopters; fourteen Piaggio P.149D, eight L-29 Delfin, MiG-15UTI and some BAC Jet Provost trainers.

NORWAY

Royal Norwegian Air Force *Kongelige Norske Luftorsvaret*

In 1912 both the Royal Norwegian Army and the Royal Norwegian Navy were given aircraft, a Maurice Farman and a Taube respectively. Official support for military and naval aviation was granted in 1915, and the Haerens Flyvapen (Army

Air Force) and Marinens Flyvevaesen (Naval Air Service) were formed, each with its own aircraft factory to overcome the World War I shortage of aircraft for non-combatant nations. Initially the aircraft produced were Maurice Farman designs, but later Bristol F.2B fighters and Hansa Brandenburg W.33 seaplane fighters were also produced.

During the inter-war period a variety of aircraft entered service with both air arms. Both services received the nationally-designed MF.9 fighter, MF.11 reconnaissance, MF.8 and MF.10 training seaplanes. The Naval Air Service also received Douglas DT-2B torpedo-biplanes and Heinkel He 115 seaplanes, while the Army received some thirty Curtiss Hawk 75A and Gloster Gladiator fighters, Caproni Ca.310 and Ca.312 bombers, Douglas DB-8A attack aircraft, and Fokker C.V and C.V.D reconnaissance-bombers. Peacetime strength had been put at thirty-six fighters and a similar number of bombers for the Army, and twenty fighters, twenty torpedo-bombers and twenty-four reconnaissance aircraft for the Navy. The small numbers of aircraft in service, and their age, meant that although a valiant and spirited resistance was mounted against the invading German forces in the spring of 1940, the outcome was a foregone conclusion. A number of aircraft and personnel flew to the United Kingdom.

The Naval Air Service re-established itself in Canada with Northrop N-3PB seaplanes, which had been ordered prior to the invasion of Norway, and eventually operated these as a part of RAF Coastal Command from British bases. The N-3PB were later replaced by Consolidated PBY-5 Catalina amphibians and Short Sunderland flying-boats. The Army Air Force personnel who had reached the UK were formed into two fighter squadrons, initially operating Hawker Hurricanes, but later Supermarine Spitfires. One of these squadrons became one of the highest scoring of Allied air force units during World War II, while also enjoying the distinction of the lowest accident rate.

In 1944 the Army Air Force and the Naval Air Service were amalgamated to form a separate service, the Kongelige Norske Luftforsvaret (or Royal Norwegian Air Force), with an initial strength of three fighter squadrons, two bomber, one reconnaissance and one transport squadrons. Equipment consisted of

Supermarine Spitfire IXs, de Havilland Mosquito VIs, Consolidated PBY-5 Catalinas, Airspeed Oxfords, Avro Ansons, ex-BOAC and RAF Lockheed Lodestars, Fairchild PT-26 and North American T-6 Harvards.

A Royal Commission in 1949 proposed a strength of eight interceptor, one night-fighter, two photo-reconnaissance, one bomber and one transport squadrons, each with eight aircraft. The first jets, de Havilland Vampire Mk.3s, had been acquired in 1948, and a further twenty-five were ordered in 1949. In the meantime Norway became a member of the North Atlantic Treaty Organization, and started to receive military aid from the United States; initially this took the form of two hundred Republic F-84E Thunderjets to form eight KNL fighter-bomber squadrons. In 1956, one of these squadrons became a reconnaissance squadron operating Republic RF-84F Thunderflash aircraft, while in 1957, North American F-86K and F-86F Sabres started to replace the Thunderjets. During this period, Douglas C-47 Dakota and Fairchild C-119F Packet transports entered service, with CCF Norseman and de Havilland Canada DHC-3 Otter light transport and communications aircraft; while Bell 47D/G Sioux helicopters and Grumman HU-16 Albatross amphibians also appeared.

In 1963, Lockheed F-104G Starfighters entered service, and during the late 1960s, Northrop F-5As were acquired. The KNL is still noted for its high standard of safety.

Currently the KNL, which has some 9,000 men, operates one squadron with twenty Lockheed F-104G Starfighters; four squadrons each with sixteen Northrop F-5A fighter-bombers and one with sixteen RF-5A reconnaissance-fighters; six Lockheed P-3B Orion maritime-reconnaissance aircraft and eight Grumman HU-16 Albatross amphibians; six Lockheed C-130H Hercules, ten Douglas C-47 Dakota and four de Havilland Canada DHC-6 Twin Otter transports; thirty-two Bell UH-1H Iroquois, five Bell 47D Sioux and six Agusta-Bell 47J, and four Sikorsky UH-19 helicopters; with SAAB-91 Safir and Lockheed T-33A trainers; Cessna O-1E Bird Dog and Piper L-21 AOP and liaison aircraft. Four Nike-Ajax and Nike-Hercules surface-to-air missile battalions are deployed. Ten Westland SH-3D Sea King anti-submarine helicopters were delivered in 1971.

PAKISTAN

Pakistan Air Force

The partition of India in 1947 led to the formation of the two independent states of India and Pakistan, the latter being divided into two, East and West Pakistan, separated by 1,100 miles of Indian territory. Initial equipment of the Royal Pakistan Air Force (as the service was known until Pakistan became a republic within the British Commonwealth in 1956) consisted of two former Royal Indian Air Force squadrons equipped with Hawker Tempest fighters and Douglas C-47 Dakota transports, with de Havilland Tiger Moth and North American T-6 Harvard trainers. RAF assistance was given during the early years, and this included the secondment of RAF personnel to the RPAF.

Immediately after the RPAF came into existence, moves were made to build it up to a strength of three Hawker Tempest fighter squadrons, a Handley Page Halifax 9 bomber squadron, a Douglas C-47 Dakota transport squadron, and a communications squadron with Auster AOP 5s, a Dakota, North American Harvards, a Vickers Viking and two de Havilland Doves. In 1950 the first of a batch of Hawker Fury fighters entered service, at the same time as the first of an order for sixty-two Bristol 170M Freighter transports, intended for the maintenance of communications between the two halves of Pakistan. During the early 1950s these aircraft were followed by thirty-six Vickers Attacker FB.2 fighter-bombers, Pakistan's first jets.

In 1954, Pakistan became a member of the South-East Asia Treaty Organization and began to receive military aid from the United States. Later, Pakistan joined the Baghdad Pact, which is now known as the Central Treaty Organization. Martin B-57B (licence-built Canberras) were placed in service in 1958, at the same time as a VIP transport Vickers Viscount turboprop airliner, and supplemented North American F-86F Sabre jet fighters which had started to arrive in Pakistan in 1956. Lockheed F-104A Starfighter interceptors were supplied by the United States in 1962, but American military aid was suspended in 1965 as a result of territorial disputes between Pakistan and India over Kashmir and the Rann of Kutch, which led to aerial and ground warfare on a couple of occasions. During the late 1960s, Pakistan received military equipment from China, including Shenyang F.6s (Chinese-built MiG-19s) and Ilyushin Il-28s. The PAF's Sabres fared badly in battles with the IAF's Gnats.

Currently the PAF, which has 15,000 men, operates one squadron with ten Lockheed F-104A Starfighter interceptors; seven squadrons with a total of 112 North American F-86F Sabre, five squadrons with a total of eighty Shenyang F.6 and one squadron with fifteen Dassault Mirage IIIE and three IIIR fighter-bombers; two squadrons with thirty-two Martin B-57B Canberra and one squadron with sixteen Ilyushin Il-28 light jet bombers; one squadron with five Lockheed T-33A tactical reconnaissance aircraft; a transport squadron operating nine Lochkeed C-130B Hercules, and a few Douglas C-47 Dakotas; ten Sikorsky UH-19, eight Sud Alouette III, two Mil Mi-6, four Kaman HH-43B Huskie and three Bell 47G Sioux helicopters; six Dassault Mirage IIID, Mikoyan MiG-15UTI, Cessna T-37B and Lockheed T-33A, and North American T-6 Harvard trainers. Four Grumman HU-16 Albatross amphibians are employed on search and rescue duties. Thirty Dassault Mirage 5 fighter-bombers are on order for 1972 delivery and these will start Sabre replacement.

Pakistan Army Aviation

The Pakistan Army operates ninety Bell 47G Sioux helicopters, Beech L-23s and Cessna O-1E Bird Dog aircraft on AOP, liaison and communications duties.

Pakistan Navy

The Pakistan Navy operates a small number of Sikorsky H-19 helicopters on rescue and light transport duties.

PARAGUAY

Paraguayan Air Force *Fuerza Aérera del Paraguay*

During the 1930s the Paraguayan Army operated Fiat CR.30 fighters and Caproni Ca.101 bombers in a war against Bolivia, backed by Breda Ba.25 trainers and later supplemented by Fiat CR.32 fighters. Although Paraguay won the war, and in 1938 some Breda Ba.65 strike aircraft were bought, deliveries since then have been entirely of transport, liaison and training types. During the 1940s, Vultee Valiant, Fairchild M-62, North American NA-16 and Boeing-Stearman PT-17 Kaydet trainers were in service with a number of Beech C-45 and Douglas C-47 Dakota transports. In 1948, Paraguay was a founder-member of the Organization of American States.

Currently the Fuerza Aérera del Paraguay operates ten Douglas C-47 Dakota, two C-54, a Convair 240 and a de Havilland Canada DHC-6 Twin Otter transports; North American T-6G Texan and Fairchild M-62 trainers; a Grumman JRF Goose amphibian; ten Bell 47G Sioux and three Hiller UH-12E helicopters. Six of the Texans are armed.

PERU

Peruvian Air Force *Fuerza Aérea Peruana*

The Peruvian Government acquired twelve Avro 504K trainers, a Curtiss J.N.4 and four Morane-Saulnier Parasols in 1920 to form a national air arm, and followed this with two Spad S-7C fighters, a Blackburn Kangaroo bomber, and three Bristol F.2B fighters, marking the beginning of Peruvian military aviation That same year, the United States provided assistance in forming a Peruvian Naval Air Service, with two Curtiss Seagull flying-boats. D.H.9 bombers and Vought UO-Q Corsair observation aircraft followed.

In 1929 the two air arms were amalgamated to form the Cuerpo de Aeronáutica del Perú, which possessed Nieuport 121C and Curtiss Hawk fighters; Douglas M-4 seaplane bombers;

Vought UO-Q and O2U-1E Corsair and Potez 39A AOP aircraft; Boeing 40B transport aircraft; Avro 504R Gosport, Stearman C-3R, Hanriot 240 and Morane-Saulnier M.S.110 trainers. During the early 1930s, Fairey Fox II bombers and Gordon general-purpose aircraft were introduced. The CAP took part in a number of operations during a border dispute between Peru and Colombia in 1933.

An Italian Air Mission which visited Peru in 1935 resulted in the purchase of Capronia Ca.114 fighters, Ca.135 bombers, Ca.111 transports, Ca.310 general-purpose aircraft and Ca.100 trainers. An attempt to establish a Caproni factory in Peru failed, although some of the aircraft ordered were assembled in Peru. During the period immediately before World War II broke out in Europe, Curtiss Hawk 75-A and North American NA-50A fighters, Douglas DB-8A bombers, Vultee 54 and Curtiss-Wright CW-22 trainers were obtained, along with a few Faucett F-19 transports. Later attempts to buy aircraft were unsuccessful because of the commitment of the world's aircraft industries to the war effort. In 1941 the CAP participated in a number of border clashes with neighbouring Ecuador, Peru successfully repelling the invading Ecuadoran forces.

In 1948, Peru became a member of the Organization of American States, and afterwards started to receive American military aid, including surplus USAF aircraft. Initial assistance included twenty Republic F-47D Thunderbolt fighter-bombers; twenty North American B-25J Mitchell bombers; and a dozen or so Lockheed PV-2 Harpoon maritime-reconnaissance aircraft; Consolidated PBY-5A Catalina amphibians; Curtiss C-46 Commando and Douglas C-47 Dakota transports; Beech T-11 Kansan, North American T-6 Texan, Boeing-Stearman PT-17 Kaydet and Fairchild PT-26 trainers; followed in 1949 by four de Havilland Canada DHC-2 Beaver transports bought from Canada. A number of pre-war aircraft remained in service until the mid-1950s.

The CAP became the Fuerza Aérea Peruana, (or Peruvian Air Force) in 1950. No further aircraft orders were made until 1955 when the first jets, North American F-86F Sabre fighters and Lockheed T-33A trainers entered service, followed in 1956 by sixteen Hawker Hunter F.52 fighter-bombers and eight

English Electric Canberra B.8 bombers, which supplemented eight Douglas B-26 Invaders delivered the previous year. Additional Canberras, many of them second-hand, have been delivered since, along with Lockheed F-80C Shooting Star fighters. Currently the FAP is South America's best equipped and trained air force. The most recent additions to its strength have been Dassault Mirage 5 fighters ordered from France after the United States requested Great Britain not to sell BAC Lightnings to Peru.

The present strength of the FAP is 9,000 men, and it operates twenty North American F-86F Sabre and ten Lockheed F-80C Shooting Star fighters; fourteen Dassault Mirage 5; sixteen Hawker Hunter F.52 and ten Republic F-47D Thunderbolt fighter-bombers; eight Douglas B-26 Invader and twenty-two BAC Canberra B(1)8 and B.2 bombers; eight Lockheed T-33A armed-trainers; six Lockheed PV-2 Harpoon maritime-reconnaissance aircraft; five Grumman HU-16 Albatross amphibians; nine de Havilland Canada DHC-2 Beaver, three DHC-6 Twin Otter and sixteen DHC-5 Buffalo, six Lockheed C-130 Hercules, and eighteen Beech Queen Air transports; four Bell 47G Sioux, six Sud Alouette II and four Alouette III, nine Bell UH-1H Iroquois and Hiller UH-12E helicopters; two Hawker Hunter T.62, two Dassault Mirage M-5, fifteen North American T-6G Texan, six Beech T-34 Mentor, twenty-six Cessna T-37B and twenty-five T-41A trainers.

Peruvian Naval Aviation *Servicio Aeronavale*

Eight Bell 47G Sioux helicopters are operated on liaison and communications duties.

PHILIPPINES

Philippine Air Force

In 1935 the Philippine Constabulary formed an aviation branch equipped with Stearman Model 76 AOP aircraft and Curtiss JN-4 trainers to assist ground forces in detecting bandit hideouts. United States assistance was provided in 1940 for the aviation branch to become a Philippine Army Air Force, which was given twelve Boeing P-26 fighters as equipment, but most of these were destroyed on the ground by Japanese air attack. After this, the Philippines forces were absorbed into those of the United States.

The Philippine Republic was formed in 1946, and in the following year the Philippine Air Force was formed as a separate service. The first aircraft were North American F-51D Mustang fighter-bombers and Douglas C-47 Dakota transports, used primarily on counter-insurgency duties. A defence treaty with the United States in 1951, followed in 1955 by the Philippines becoming a member of the South-East Asia Treaty Organization, opened the door to further United States aid. Replacement of the Mustangs with North American F-86F Sabres in 1957 was followed in 1959 by the arrival of Japanese-built Beech Mentor trainers. A number of supporting aircraft were also placed in service, including Grumman HU-16A Albatross amphibians, Sikorsky H-19A and H-34 and Hiller FH-1100 helicopters, Lockheed T-33A, North American T-6G Texan and T-28 Trojan trainers. One squadron of Sabres was replaced by Northrop F-5A Freedom Fighters during the late 1960s.

Currently the Philippine Air Force, which has some 9,000 men, operates one squadron of twenty Northrop F-5As, and two squadrons with a total of thirty North American F-86F Sabres, on ground-attack and fighter-bomber duties; ten Douglas C-47 Dakota, four NAMC YS-11 and three Fokker F-27 Troopship transports; a communications Aero Commander 500; some Grumman HU-16A Albatross amphibians; Stearman L-5 AOP aircraft; two Mitsubishi-Sikorsky S-62A, two Sikorsky H-34 and five H-19, a Bell 47D Sioux and eight Hiller FH-1100 helicopters; North American T-6G Texan and T-28 Trojan, thirty-six Fuji-Beech T-34 Mentor and Lockheed T-33A trainers.

POLAND

Polish Air Force *Polskie Lotnictwo Wojskowe*

Prior to 1918, Poland had been occupied by successive foreign powers, and independence was only gained with the ending of

World War I. Although there were Polish personnel in the Russian, German and Austrian forces, the history of Polish military aviation did not start formally until 1919 when a seven-squadron air force was formed as a part of the Army, operating Spad S.7C fighters, Breguet Br.14A.2 reconnaissance and Br.14B.2 bomber aircraft, and Salmson SAL.2-A2 reconnaissance-bombers: in total almost a hundred aircraft. A number of local volunteer flying units were absorbed into this force with their aircraft during the year or so which followed, by the end of which time it operated most types of military and civil aircraft then available. Additional Bristol F.2B fighters, Sopwith Dolphins and Martinsyde F.4s were bought from Great Britain, while a number of D.H.9 bombers and Sopwith Camel fighters were presented as a gift, and a number of former German and Italian aircraft were also purchased. The purpose of all of this activity was to build up the Polish Air Force for a war with the newly-formed USSR, which lasted from late 1919 for a year, and during which the Polish Air Force effectively saved Poland from defeat.

Between the wars, attempts were made to establish a national aircraft industry with both private and state-owned concerns producing national designs as well as licence-building foreign aircraft, notably Fokker F.VII transports and Avia B.H.33 trainers. The more significant Polish aircraft included the PZL P-1, P-6, P-7 and P-11 fighters during the early 1930s, along with P-23B reconnaissance-bombers and, just before World War II, the PZL P-37 bomber. There were also RWD-8, RWD-14 and Lublin R-XIII army co-operation aircraft and PWS-2 trainers.

In 1938 the Polish Air Force became a separate service after a reorganization of the Army. The newly independent service had fifty-five fighters, seventy-six bombers, plus transport, liaison and training aircraft. The Army had 260 aircraft attached to it for support, liaison and AOP duties. Elaborate and ambitious plans were laid for expansion of the force immediately before World War II started, but most of the aircraft were some ten years behind the times, and only the PZL-37 bomber was in service at the outbreak of war out of a number of more modern aircraft planned.

On 1 September 1939, German forces invaded Poland and, in spite of the Polish Air Force's efforts, the overwhelming strength and superiority of the German forces and their equipment made defeat inevitable. Polish personnel took the few surviving aircraft to Rumania, where they left the aircraft and managed to get to France. Polish personnel flew Morane-Saulnier M.S.40C and Caudron C.714 fighters before the fall of France, after which they went to the United Kingdom. Polish squadrons in the Royal Air Force were equipped with Hawker Hurricane fighters and Fairey Battle bombers, replaced later by Supermarine Spitfires and Vickers Wellingtons respectively. One squadron operated Boulton-Paul Defiant night-fighters, which were replaced by Bristol Beaufighters and then de Havilland Mosquitoes. Before the war ended, Polish personnel had flown North American F-51D Mustang fighters, North American B-25 Mitchell, Handley Page Halifax and Consolidated B-24 Liberator bombers, while some Polish squadrons were also formed in Russia. After the liberation of Poland in 1944, few ex-RAF personnel returned to Poland, and of those who did none were allowed to stay in the new Polish Air Force for any length of time before being sent to detention camps.

The new Polish Air Force used its former Russian wartime aircraft at first, including Polikarpov Po-2, Yakovlev Yak-1, Yak-3 and Yak-9 fighters; Ilyushin Il-2 bombers; and Polikarpov Po-2, Yakovlev UT-2 and Petlyakov Pe-2 trainers. The Yak-9 became the standard fighter, while Ilyushin Il-10 ground-attack aircraft and Tupolev Tu-2 bombers were introduced. Lisunov Li-2 (C-47) transports and Yak-11 and Yak-18 trainers were introduced later, while in 1950 the first jets, the Yak-23 fighters, were placed in service. These were followed during the early 1950s by Mikoyan MiG-15 fighters, including the Polish-built version, the LIM-2. Nationally-designed Junak-3 trainers also entered service, along with Yak-17 jet trainers, Ilyushin Il-28 jet bombers and Mil Mi-1 helicopters, of which the Polish-built version was the SM-1.

In 1957, MiG-17 fighters started to supplement and then replace the MiG-15s and Yak-23s while in turn these have been largely superseded by MiG-19 and MiG-21 fighters and interceptors respectively, while Sukhoi Su-7s have taken over the ground-attack duties.

Currently the PLW has a personnel strength of some 25,000, and operates forty-five Mikoyan MiG-21, MiG-19 and MiG-17 squadrons in the interceptor role; twelve Sukhoi Su-7 ground-attack squadrons; three MiG-17 reconnaissance squadrons and six Ilyushin Il-28 bomber squadrons, with twelve aircraft per squadron, a total of forty to fifty transport aircraft, including Antonov An-2 and An-12, Ilyushin Il-12, Il-14 and Il-18, and Lisunov Li-2(C-47) types, and about forty Mil Mi-1, Mi-4 and Mi-6 helicopters, with three hundred trainers of Mikoyan MiG-15UTI, MiG-21UTI, Sukhoi SU-7U, Yakovlev Yak-11 and Yak-18, Ilyushin Il-28U and Junak 3 types.

PORTUGAL

Portuguese Air Force *Forca Aérea Portuguesa*

Portuguese military aviation history dates from 1917, when the Arma da Aeronautica and the Aviação Maritima were both formed, although Portuguese Army and Navy officers had received flying training in the United Kingdom and France, and there had been a Portuguese flying school in existence for a few years. The Arma's first aircraft were Spad S-7C fighters and Breguet Br.14A.2 bombers, while the Navy operated Fairey Campania seaplanes and Short Felixstowe F.3 flying-boats, to which Fairey III seaplanes were later added. In 1924 the Army's air arm was operating some twenty-five aircraft, and its future strength was put at three squadrons – fighter, bomber and reconnaissance – while the equipment of the period included Martinsyde F.4 Buzzard fighters, Caudron G.III and Avro 504K trainers. Three years later saw the Navy with three Fokker T.IV reconnaissance-seaplanes, three H.S.2L and seven C.A.M.S. 37 flying-boats and five Hanriot H.41 training seaplanes.

The 1930s saw the early steady progress continue, and by 1935 licence-built Potez XXV bombers and Vickers Valparaiso reconnaissance aircraft, of which there were sixteen and twenty respectively, plus an assortment of training types, including twenty-two de Havilland Tiger Moths, were in service, although the Breguet Br.14A.2s were still soldiering on. Later the Army and the Navy both received licence-built Morane-Saulnier M.S.233 and Avro 626 trainers, and after a reorganization the Army received Hawker Hind and thirty Gloster Gladiator fighters, Breda Ba.65 ground-attack aircraft and Junkers Ju 86K bombers, replacing much of the earlier equipment. A few Supermarine Spitfire I fighters were also obtained later.

Although neutral during the war, Portugal retained very cordial relations with the Allies. Apart from the Spitfires already mentioned, the Arma also received Hawker Hurricane, Bell Airacobra and Curtiss Hawk 75A fighters; Bristol Blenheim and Consolidated B-24 Liberator bombers; Miles Master, Magisters and Martinets, and Airspeed Oxfords for training. The Aviação Maritima received Bristol Beaufort and Blenheim, Short Sunderland flying boats, Grumman G-21 amphibians, and Lockheed Hudson bombers. From 1943 onwards, RAF Coastal Command aircraft operated from Santa Maria in the Canaries with the full permission of Portugal.

Post-war re-equipment was long delayed, only twenty or so Republic F-47D Thunderbolt fighter-bombers and a handful of Douglas C-47 Dakota and C-54 Skymaster transports arriving during the late 1940s, although Portugal joined the North Atlantic Treaty Organization. In 1957 the Naval and Army air arms were amalgamated, forming the Forca Aérea Portuguesa, a separate service which was blessed with Portugal's first jets, Republic F-84G Thunderjets, in 1958 to replace its Spitfires and Hurricanes; while, for training, Lockheed T-33A jets and de Havilland Canada Chipmunks (built in Portugal) were bought, and fifteen ex-Royal Navy North American T-6 Harvards were provided as a gift. Other aircraft of this period included Boeing SB-17G Fortress bombers, Lockheed PV-2 Harpoon maritime-reconniassance aircraft, Grumman SA-16A Albatross amphibians, and Sud Alouette II and Sikorsky H-19A Chickasaw helicopters. North American F-86F Sabre fighters arrived later, but in recent years Portugal has faced considerable difficulty in obtaining military equipment because of opposition to her maintaining a number of African colonies – where FAP units are based in support of ground forces.

Currently the FAP has some 17,500 men and operates one squadron of twenty-five Republic F-84G Thunderjet, two with a total of fifty North American F-86F Sabre, and two with a total of thirty-six Fiat G.91R-4 fighter-bombers; two bomber squadrons of twelve aircraft each, one with Douglas B-26 Invaders and one of Lockheed PV-2 Harpoons; a maritime-reconnaissance squadron operates twelve Lockheed P-2E Neptunes. There are twenty Nord Noratlas, fifteen Beech C-45, forty Douglas C-47 Dakota, five C-54 Skymaster, and a few DC-6 transports; eighty Sud Alouette III and seven Alouette II, twelve SA.330 Puma and a number of Sikorsky UH-19A Chickasaw helicopters; North American T-6 Harvard, fifteen Lockheed T-33 and thirty Cessna T-37C, a few de Havilland Vampire T.55 and de Havilland Canada Chipmink trainers; and twenty-five Dornier Do 27 liaison aircraft.

RHODESIA

Rhodesian Air Force

The history of Rhodesian military aviation goes back to before World War II when the Government of the then Southern Rhodesia offered to provide three squadrons for the Royal Air Force, as well as participating in the Empire Flying Training Scheme. The three Rhodesian squadrons in the RAF operated Supermarine Spitfire and Hawker Typhoon fighters, and Avro Lancaster bombers. Flying training for the RAF took place in Rhodesia for many years after the war, due to the excellent flying conditions prevailing in the area. The post-war period also saw the formation of the Southern Rhodesian Air Force, which had one squadron operating Douglas C-47 Dakotas, an Avro Anson and a de Havilland Rapide with some Leopard Moths.

In 1951 the SRAF became fully operational with eleven Supermarine Spitfire 22 fighters, replaced in 1953 by de Havilland Vampire FB.9 fighter-bombers, and two years later Vampire T.11 trainers were delivered. A squadron of English Electric Canberra jet bombers was formed in 1958, followed by a Hawker Hunter fighter-bomber squadron. Some Hunter T.52s were also delivered.

Southern Rhodesia was part of the Federation of Rhodesia and Nyasaland from 1953 until 1963, during which time the SRAF became the Royal Rhodesian Air Force, a title retained by the air force of Southern Rhodesia after the Federation was dissolved. Independence for Northern Rhodesia and Nyasaland was coupled with changes of name to Zambia and Malawi respectively, thus allowing the 'Southern' to be dropped from the name of the remaining state. In 1965 a Unilateral Declaration of Independence was made to prevent any grant of independence on terms unacceptable to Rhodesians. This led to United Nations sanctions against Rhodesia, which seem largely to have failed, and in 1969 to the declaration that Rhodesia was a republic, so dropping the 'Royal' prefix from the Rhodesian Air Force's title.

Currently the RhAF has some 1,200 men and operates one squadron of twelve BAC Canberra B.2 bombers; one squadron of twelve Hawker Hunter FGA.9 and one squadron of twelve de Havilland Vampire FB.9 fighter-bombers; a squadron of twelve BAC Provost T.52 armed-trainers; four Douglas C-47 Dakota transports; and eight Sud Alouette III helicopters; with Hawker Hunter T.52, and de Havilland Vampire T.55 trainers; while a Beechcraft Baron was obtained in 1967 for communications duties.

RUMANIA

Rumanian Air Force

The Rumanian Army formed a Flying Corps in 1910, and by the end of 1911 this had four Blériots and four Henri Farmans, which were augmented by a small number of Bristols and two Morane Type Fs in 1912. Apart from the delivery of a small number of Nieuport 12 and 17 aircraft, no further progress was made, and this small force was quickly defeated by the Central Powers at the start of World War I.

After the war ended, a Directorate of Army Aviation was established, with a strength of three groups, each with three squadrons. A fighter group operated Spad S-7C.1s, a bomber group D.H.9s and Breguet Br.14.Bs, while Breguet Br.14A.2

and Brandenburgs operated in the reconnaissance group – a total of seventy-two aircraft. During the late 1920s seventy Armstrong-Whitworth Siskin III and sixty Spad S-61C.1 fighters, 120 Potez XXV and thirty XXVII reconnaissance aircraft, numbers of Breguet Br.19B.2 bombers and Savoia S.59 flying-boats, with Morane-Saulnier M.S.35 trainers, were obtained – many of the Potez and Morane-Saulnier aircraft being assembled in Rumania.

The early 1930s saw the nationally-designed S.E.T. XV fighter, S.E.T. 7K reconnaissance aircraft, S.E.T. 7 and S.E.T. X trainers enter service, followed by fifty PZL P-11b fighters and twenty Consolidated Fleet Model 10G trainers. In 1936, Miles Hawk and Nighthawk trainers were bought, and the nationally-designed I.A.R. 37, 38 and 39 light bombers entered service. Britain supplied Hawker Hurricane fighters and Bristol Blenheim 1 bombers in 1939.

Rumania sided with the Axis powers during World War II, and received from Germany some Messerschmitt Bf 109E and Heinkel He 112B fighters, Heinkel He 111H bombers and Junkers Ju 87D Stuka dive-bombers; Heinkel He 114 reconnaissance seaplanes and Fieseler Fi 156C liaison aircraft; while licence-built Savoia-Marchetti S.M.79 bombers entered service with the nationally-designed I.A.P. 80 fighter. Rumanians fought alongside German forces on the Russian front. Towards the end of the war, Junkers Ju-88A bombers and Henschel Hs 129A-O ground-attack aircraft were supplied, but by the end of August 1944 Russian forces had invaded Rumania.

The immediate post-war period saw a newly-formed Rumanian Air Force operating its wartime equipment, before the arrival of Yakovlev Yak-9 fighters. The 1947 Peace Treaty restricted Rumania to 150 military aircraft and 8,000 air force personnel, but Russian military aid soon ensured that these limits were exceeded. In 1953 the first jets, Mikoyan MiG-15 fighters, were supplied, with Yakovlev Yak-11 and Yak-18 trainers, Ilyushin Il-10 ground-attack aircraft and Lisunov Li-2 (C-47) transports. The mid-1950s saw MiG-17 fighters enter service, and the first helicopters, Mil Mi-4s, and also Ilyushin Il-28 jet bombers. Antonov An-2 transports and MiG-19 and MiG-21 fighters and interceptors are amongst the more recent additions.

Currently the Rumanian Air Force has some 8,000 men, and operates eighteen squadrons with MiG-21, MiG-19 and MiG-17 aircraft in the interceptor and fighter-bomber roles, and two bomber squadrons with Ilyushin Il-28s – about 250 combat aircraft altogether. There are two transport squadrons with Ilyushin Il-12s and Il-14s, and Lisunov Li-2s; with ten Mil Mi-4 helicopters; and Yak-11 and Yak-18, MiG-15UTI and L-29 Delfin trainers.

SAUDI ARABIA

Royal Saudi Air Force

Saudi Arabia is a union of the Nejd, recaptured in 1913 by King Ibn Saud, uncle of the present monarch, and the Hejaz, captured in 1926. The present national title dates from 1926. The first military aircraft were some D.H.9 bombers provided by the British Government to assist King Saud in suppressing dissident tribesmen, followed in 1931 by four ex-RAF Westland Wapiti general-purpose biplanes which were operated by Britons. Italian aid took over in 1937 but, although this included a few aircraft, the next significant step was not until 1950, when a British mission arrived to reorganize what had become the Royal Saudi Air Force, and provided Avro Anson transports and de Havilland Tiger Moth trainers.

Development of the RSAF really got under way in 1952 when an American mission provided ten Temco TE-1A Buckaroo and some North American T-6 trainers, followed by the equally inevitable Douglas C-47 Dakota transports. The following year, eighteen Saudi pilots were trained in the United Kingdom by Airwork and by Air Services Training, while a few pilots were also trained in Egypt under a Saudi-Egyptian defence agreement which resulted, too, in four de Havilland Vampire FB.52 fighter-bombers in 1957 before the agreement ended. In the meantime the RSAF had also received nine Douglas B-26 Invader bombers and a number of de Havilland Canada Chipmunk trainers. The late 1950s and early 1960s saw other American equipment entering service in the form of sixteen North American F-86F

Sabre fighter-bombers, Lockheed T-33 and Beech T-34 Mentor trainers, six Fairchild C-123 Provider and four Lockheed C-130E Hercules transports. At one time, a Westland Wigeon helicopter was operated as a VIP transport, as was an ex-Royal Jordanian Air Force Vickers Varsity.

In 1966 an order for thirty-four BAC Lightning F.53 'multi-mission' aircraft and six Lightning T.55 and twenty-four BAC 167 Strikemaster trainers was placed as part of an integrated air defence system designed to combat possible Egyptian, Syrian or Iraqi attacks. This has been brought to operational readiness by Airwork Services, a British firm.

Currently the RSAF has a strength of some 5,000 men, and operates thirty-four BAC Lightning F.53s in the strike role; sixteen North American F-86F Sabre fighter-bombers; twenty-four BAC 167 Strikemaster armed-trainers; eight Douglas C-47 Dakota and two C-118 Liftmaster, six Fairchild C-123 Provider and nine Lockheed C-130E Hercules transports; two Sud Alouette III, two Agusta-Bell 204, twenty-four 205 and ten 206 helicopters; Hawker Hunter, BAC Lightning T.55 and Cessna T-41A trainers; with BAC Thunderbird and also Hawk surface-to-air missiles.

SINGAPORE

Singapore Air Defence Command

Following Singapore's withdrawal from the Federation of Malaysia, Singapore set up her own defence forces, although strong defence links with Malaysia are maintained. Current equipment of the three hundred-strong SADC is sixteen BAC 167 Strikemaster armed-trainers and twenty Hawker Hunter FGA.9 ground-attack aircraft, eight Sud Alouette III helicopters and a similar number of Cessna 172 AOP and liaison aircraft. More sophisticated aircraft may be ordered in the future, with the Dassault Mirage F.1 interceptor or G.4 swing-wing bomber as possibilities.

SOMALI

Somalian Aeronautical Corps   *Cuerpo Aeronautica del Somalia*

The former British Somaliland Protectorate became independent in 1960, at about the same time as the Italian Trust Territory of Somalia, and an Italian-sponsored Air Corps was left in being as a basis for a national air arm. Since this time, development has been largely through Soviet aid. Currently the CAS is some 2,000 men strong, and operates twelve Mikoyan MiG-15 and six MiG-17 fighter-bombers; six Beech C-45, a Douglas C-47 Dakota, and an Antonov An-24 as transports; twenty Yakovlev Yak-11, ten Piaggio P.148, and six MiG-15UTI trainers.

SOUTH AFRICA

South African Air Force   *Suid Afrikaanse Lugmag*

South Africa's history of military aviation began in 1915 with the formation of the South African Aviation Corps, although some South African Army officers had received flying training in 1913 and many South Africans were already serving in the Royal Flying Corps. Initial equipment of the SAAC consisted of ex-RNAS Henri Farmans and B.E.2as, and the force was intended to operate against German South-West Africa, but there was in fact little air activity on that front. However, the SAAC did fight with some distinction against German forces in East Africa. In 1918 the SAAC was disbanded.

The South African Air Force was formed in 1920 as a separate service, using the 'Imperial Gift' of one hundred ex-RAF aircraft provided by the British Government, including S.E.5A fighters, forty-eight D.H.4, D.H.9 and D.H.9A bombers, and a few Avro 504K trainers. During the early years, there were a few police actions, and in 1925 an air-mail service was operated from Durban to Cape Town, while surveying and training of Citizen Defence Force pilots, along with a few other duties, kept the SAAF well occupied. In 1929 twenty Avro Avian trainers were purchased to replace the 504Ks, while thirty-one Westland Wapiti general-purpose biplanes were also placed in service at about this time, and another twenty-seven of this type assembled locally. Sixty Avro Tutors were licence-built for the SAAF during the early 1930s.

A major expansion programme was started in 1936. Sixty-five Hawker Hind bombers were manufactured under licence in South Africa as the Hartebeeste, while equipment supplied direct from the United Kingdom included Hawker Fury and Hurricane and Gloster Gladiator fighters, Fairey Battle and Bristol Blenheim bombers. Some one hundred combat aircraft were in service at the outbreak of World War II in September 1939, and in addition Junkers Ju 86 airliners of South African Airways were pressed into service on maritime-reconnaissance duties.

Although South Africa was not a part of the Empire Flying Training Scheme, SAAF training facilities were considerably expanded and placed at the disposal of the RAF. Throughout the war the SAAF operated maritime-reconnaissance patrols in the South Atlantic and the Indian Ocean, while SAAF units fought in the Western Desert in North Africa against Italian and German forces; in East Africa protecting Kenya and the Sudan from Italian forces, and eventually liberating Ethiopia; and in Europe after taking part in the Allied invasion of Sicily. Aircraft operated by the SAAF during World War II included Curtiss Kittyhawk, Mohawk and Tomahawk, Hawker Hurricane, Supermarine Spitfire and North American Mustang fighters; Martin Maryland, Marauder and Baltimore, Vickers Wellington, Bristol Beaufort and Blenheim, Douglas Boston and Consolidated Liberator bombers; Vickers Warwick, Lockheed Ventura, Lodestar and Harpoon maritime-reconnaissance aircraft, with Short Sunderland flying-boats and Consolidated PBY-5 Catalina amphibians also operating in this role; Vickers Valentia, Junkers Ju 52/3M and Douglas Dakota transports; de Havilland Tiger Moth, Airspeed Oxford, Avro Anson, North American Harvard, Northrop Nomad, Miles Master and Hawker Audax trainers. After the war the SAAF was singled out for comment by Air Chief Marshal Tedder (of the RAF) for its notable contribution to victory.

After the war was over, the SAAF was reorganized as a small force capable of rapid expansion from SAAF Reserve and Active

Citizen Force personnel. Wartime equipment lingered on for a few years, and the SAAF took part in the Berlin Airlift. In 1950 the SAAF sent a small force to fight alongside United Nations forces in Korea, whose North American F-51D Mustangs often had to face Communist Mikoyan MiG-15 jet fighters. It was in Korea, in 1952, that the SAAF received its first jets, North American F-86 Sabres on loan from the USAF until the SAAF force returned to South Africa the following year. In the meantime, de Havilland Vampire FB.5 jet fighter-bombers with Vampire T.55 trainers had started to replace the Spitfires. Canadair CL-13 Sabre 6 fighters, de Havilland Dove and Heron transports, and Sikorsky S.55 helicopters were delivered in 1956. The following year, the Short Sunderlands were replaced by Avro Shackleton MR.3 maritime-reconnaissance aircraft for defence of the all-important Cape shipping route.

During the early 1960s a limited expansion programme was started to maintain a balance of power in Africa where certain newly independent states were hostile to South Africa. Political feeling against South Africa led to withdrawal from the British Commonwealth in 1961, and, although a United Nations resolution against selling arms to South Africa excluded arms for external defence, during the middle of the decade this was interpreted by the British Government as applying to all arms. However, an Anglo-South African Defence Treaty (known as the Simonstown Agreement) remained in force and, before the total ban, sixteen Hawker Siddeley Buccaneer low-level strike bombers were supplied. After the ban, France took over the role as main supplier of arms to South Africa, although upwards of two hundred Italian Aermacchi MB.326K armed-trainers have been built in South Africa under licence recently. A change of government in Britain in 1970 and the increased importance of the Cape route to shipping through closure of the Suez Canal has led to a change in policy on the question of arms sales.

Currently the SAAF has some 8,000 men, and operates one squadron of sixteen Dassault Mirage IIICZ interceptors; one squadron of twenty Mirage IIIEZ and one of sixteen Canadair CL-13 Sabre 6 fighter-bombers; fifteen Hawker Siddeley S.50 (rocket-boosted) Buccaneer and nine BAC Canberra B.12 bombers in two squadrons; four Dassault Mirage IIIRZ reconnaissance-fighters; while a further Sabre squadron and some thirty de Havilland Vampire FB.9 fighter-bombers are in reserve. Eight Avro Shackleton MR.3 maritime-reconnaissance aircraft are in one squadron; there are nine Transall, forty Douglas C-47 Dakota and four C-54 Skymaster, seven Lockheed C-130B Hercules and a VIP Vickers Viscount as transports; nine Piaggio P.166 communications aircraft; six Sud Alouette II, fifty Alouette III, twenty SA.330 Puma and sixteen Super Frelon helicopters; while eight Westland Wasp helicopters are available for operations from South African warships. Some eighty or so Aermacchi MB.326Ks (known as the Impala) are used for training out of some 250 built in South Africa, the remainder being armed and used by the Active Citizen Force, whose Harvards they replaced. Crotale (Cactus) surface-to-air missles are deployed. A further sixteen Buccaneers and eight Hawker Siddeley Nimrod maritime-reconnaissance aircraft (to replace Canberras and Shackletons respectively) are likely to be bought from the UK during 1971. Licensed production of Dassault Mirage III and FI aircraft is now taking place. Army Air Corps operates eighteen Cessna 185 liaison aircraft.

SOUTH YEMEN REPUBLIC

Air Force of the South Yemen People's Republic

Originally formed with RAF assistance during the mid-1960s before Great Britain's withdrawal from Aden, the present title was adopted after independence. Equipment consists of four BAC 167 Strikemaster and four Jet Provost T.5 armed-trainers, a similar number of Douglas C-47 Dakota and six de Havilland Canada DHC-2 Beaver transports, and six Westland-Bell 47G Sioux helicopters. Although requested, further British aid was refused after independence due to the anti-Western line of the new government, and it is almost certain that all future aid will come from Egypt or Eastern Europe.

SPAIN

Spanish Air Force *Ejercito del Aire*

Spain's history of military aviation dates from as long ago as 1896, when a balloon company was formed as a part of the Spanish Army, and this saw action during the Riff uprisings in Morocco some fourteen years later. It was not until 1911 that the balloon company became the Aeronáutica Militar Española with an initial strength of two Henri Farman, two Maurice Farman, two Bristol and six Nieuport aircraft. During the next four years before World War I started, additional Maurice Farman and Nieuport aircraft were acquired, with some Morane-Saulnier M.S.14 and Löhner biplanes. Spain remained neutral during the war, and in common with other neutral countries found difficulty in obtaining aircraft for the AME and the newly formed Aeronáutica Navale. This led to licence-construction of D.H.4 bombers and Morane-Saulnier Parasols, while the Spanish-designed Flecha also entered production. A few Curtiss F flying-boats were also obtained for the Aeronáutica Navale.

The end of the war meant the release of vast numbers of redundant aircraft on to the market, and the two Spanish air arms benefited from the timely acquisition of Ansaldo, Bristol F.2B 'Brisfit', Martinsyde F.4A and Spad S.13C.1 fighters, Farman F-50, Salmson SAL2-A2, Breguet Br.14A-2 and D.H.4 bombers, Macchi M.9 and Savoia S.16 flying-boats, and a number of Caudron G.III trainers. Many of these aircraft saw service in Morocco, where Spanish forces were fighting the Riff until 1926. During the early 1920s additional aircraft finding their way into Spanish service included Dornier Wal and Macchi M.18 flying-boats, Blackburn Velos seaplanes, and Supermarine Scarab amphibians for the Navy; and licence-built Breguet Br.19A-2 and D.H.9 bombers, twenty Fokker C.IV AOP and C.III training aircraft, with a number of Avro 504K trainers, for the Army.

Later a number of Dornier Wals were built in Spain, while licence-built Nieuport 52C-1 fighters and the Spanish-designed Loring R-1 reconnaissance aircraft were also produced; but the late 1920s and early 1930s saw in reality few new aircraft enter service, except for these small orders and some Vickers Vildebeeste torpedo-bombers in 1931, and the strength of the two Spanish air arms dropped sharply from seven hundred to three hundred aircraft. Also in 1931 the monarchy was replaced by a republic.

In 1936 the Spanish Civil War between General Franco's Nationalists and Republican elements began, initially in Morocco. The Republicans had some two hundred aircraft against the Nationalists' sixty or so, but when Russia started to assist the Republicans, Germany and Italy came to the aid of the Nationalists. The aid given for the most part falls to be described in the German, Italian and Russian histories, particularly since aircraft provided by these nations usually had 'volunteer' pilots from the air force of the appropriate nation. However, France also supplied aircraft to the Republican forces, including one hundred Dewoitine D.373, D.500 and D.510, Lioré-Nieuport L.N.46 and Spad 510C fighters; and Potez 56 and Bloch M.B.200 bombers. Czechoslovakia supplied Letov S-231 fighters and Aero 100 general-purpose aircraft.

The war ended in 1939 with a Nationalist victory, and the new Government immediately reorganized Spanish military aviation with the Ejercito del Aire (Spanish Air Force) being formed as a separate service on the merger of the two air arms. A good indication of the foreign aid given to the two sides in the Spanish Civil War is that, even after considerable losses, there were still 1,000 aircraft available to the EdA on its formation.

When World War II started, Spain again remained neutral, and again was largely cut off from aircraft supplies. Two Spanish concerns, CASA and Hispano-Suiza, obtained licences to produce Fiat, Messerschmitt, Heinkel, Bücker, Dornier and Junkers designs. A few Messerschmitt Bf 109F fighters, Junkers Ju 88A bombers, Heinkel He 114 seaplanes and Dornier Do 24 flying-boats, with some Fieseler Fi 156 Storch general-purpose aircraft, were delivered from Germany, but for the most part Spain's Bf 109Fs and He 114s were home-produced, as were 250 He 111H bombers, one hundred Ju 52/3M transports, and Fiat C.R.32 fighters. Some Spaniards flew with the Luftwaffe on the Russian front.

Manufacture of these aircraft continued until 1953, although with modifications: the most significant was the fitting of Rolls-Royce Merlin engines (which type had been used by the Spitfires of the RAF during the Battle of Britain) to Messerschmitt Bf 109s. Although Spain is not a part of the North Atlantic Treaty Organization, a defence treaty agreed with the United States in 1953 marked the start of American military aid in return for the use of Spanish bases. In the period immediately after 1953 the EdA received two hundred North American F-86D/F Sabre fighters, fifteen Douglas C-47 Dakota transports, a number of Grumman HU-16 Albatross amphibians, Sikorsky H-19 Chickasaw and Bell 47G Sioux helicopters, thirty Lockheed T-33A and one hundred North American T-6G Texan trainers. The Spanish aircraft industry produced the CASA 201B, 207 and 352L transports and the Hispano Aviación HA-100 and HA-200 trainers.

The early 1960s saw a small batch of Lockheed F-104G Starfighters enter service, followed later by CASA-built Northrop SF-5A and SF-5B Freedom Fighters. The Spanish-American Defence Agreement was extended in 1970, and in return Spain received McDonnell Douglas F-4 Phantom fighter-bombers and a number of other aircraft. Before this, Dassault Mirage IIIEs were bought.

Currently the Ejercito del Aire has a strength of some 32,000 men, and operates twenty Lockheed F-104G Starfighters in one interceptor squadron; sixty North American F-86F Sabres remain in service; thirty-six McDonnell Douglas F-4C Phantom and thirty Dassault Mirage IIIE with seventy Northrop SF-5A fighter-bombers; twenty-five Hispano Aviación HA-220 armed-trainers; three Lockheed P-3 Orion and eleven Grumman HU-16 Albatross amphibians, operating on maritime-reconnaissance duties; two Lockheed C-130 Hercules, one hundred Douglas C-47 Dakota and C-54 Skymaster, twelve de Havilland Canada DHC-4 Caribou and a number of CASA 207 Azor transports; five Sikorsky H-19 Chickasaw, thirty Bell 47D/G Sioux and a number of Hiller UH-12C helicopters; Beech T-34 Mentor, North American T-6G Texan, Hispano HA-200 Saeta, Lockheed T-33A and Northrop F-5B trainers. Eight more helicopters, probably of American origin, are also likely to be introduced during 1971.

Naval Aviation *Marinha*

Currently, six Sikorsky SH-3D and three H-19 Chickasaw, four Westland Whirlwind, ten Agusta-Bell 47G Sioux and twelve 204B, with three Bell UH-1D Iroquois helicopters are operated, often from Spanish warships.

Army Aviation *Arma España*

Currently, twelve Bell UH-1D Iroquois and six 47G Sioux helicopters are operated on liaison and communications duties, with a number of Cessna O-1 Bird Dogs for AOP duties. Hawk surface-to-air missiles are also deployed.

SUDAN

Sudanese Air Force

The Sudan became an independent republic in 1958 after being under Anglo-Egyptian rule, and in the following year started to form the Sudanese Air Force, initially as an internal security organization operating four Hunting Provost T.51 armed-trainers, while Hunting Pembroke and Douglas C-47 Dakota transports followed.

Currently the Sudanese Air Force, which has 450 men, operates sixteen Mikoyan MiG-21s in the fighter-bomber role; eight BAC Jet Provost T.52 armed-trainers, with three Provost T.51s and five BAC 145 T.5s also in the training role; three BAC Pembroke, three Fokker F-27M Troopships and five Antonov An-24 transports.

SWEDEN

Royal Swedish Air Force *Flygvapnet*

In 1911 the Royal Swedish Navy was presented with a Blériot aircraft, while the following year the Swedish Army received a

Nieuport IVG; in each case the aircraft came from air-minded citizens. At the outbreak of World War I in 1914, the RSwN was operating the Blériot, two Henri Farmans, and a Donnet-Leveque flying-boat, while the Army Air Corps operated the Nieuport IVG, a Breguet and another two aircraft. Sweden remained neutral during the war and therefore was unable to obtain new aircraft from the hard-pressed aircraft industries of both sides. This led to licence-production of Farman F.23, Albatros C.III and Morane-Saulnier Parasol aircraft.

The return of peace in Europe saw the RSwN with about twenty-five aircraft, and the Army Aviation Corps with about twice that number. Domestic aircraft production continued, including some foreign designs, but also the Swedish-designed J.23 and J.24 fighters, S.18, S.21 and S.25 reconnaissance aircraft, and Ö.1 trainers. A number of Phoenix 122 fighters and Avro 504K trainers were also built. In 1926 the Flygvapnet (Royal Swedish Air Force) was formed to take over the RSwN and Army Aviation Corps aircraft and aircrew and to operate as a separate service. The Flygvapnet's first aircraft consisted of those post-war types already mentioned, with Nieuport 29C-1 fighters and Fokker C.V reconnaissance aircraft. Nevertheless, for the first few years of its existence the Flygvapnet was largely neglected, and it was only when the shortage of equipment reached desperate proportions that twelve Bristol Bulldog fighters, a number of ASJA J.6s, an assortment of Swedish and British-built Hawker Hart light bombers and Ospreys, twenty-five ASJA RK.26 and some forty or so de Havilland Tiger Moth trainers were obtained.

The late 1930s saw a revival of interest in military aviation as the situation in Europe pointed to the need for strengthened defences. The Flygvapnet was reorganized into eight wings: F.1, F.4, F.6 and F.7 operating bombers, F.8 fighters, F.2 naval reconnaissance, F.3 army reconnaissance, and F.5 flying training. Some sixty Gloster Gladiator fighters were built, along with forty Junkers Ju 86K and one hundred Douglas DB-8A bombers and forty North American NA-16-4 trainers.

In 1940 sixty Republic EP-1 fighters were obtained from the United States, while from Italy seventy-two Fiat C.R.42 and C.R.60, and Reggiane Re.2000 fighters, and eighty Caproni

Ca.313 bombers were obtained. Two new fighter wings, F.9 and F.10, a new reconnaissance wing, F.11, and a new bomber wing, F.12, were formed. During 1939 and 1940 a small but effective Flygvapnet unit had fought alongside Finnish forces against Russia, but Sweden was to remain neutral during World War II and, possibly in anticipation of equipment shortages, three SAAB designs, the SAAB-21 fighter, SAAB-18 bomber and SAAB-17 light bomber, were put into production, followed shortly by the SAAB-22 fighter. A number of aircraft ordered from the United Kingdom, France and the United States could not be delivered because of the war. It was in fact not until 1945, when fifty North American F-51D Mustang fighters were delivered, that outside supplies could again be relied upon.

The last SAAB-21 fighters, of which there were some three hundred, were not delivered until after the war ended, while ninety Mustangs were added to those provided just before the cessation of hostilities. The first jets, de Havilland Vampire F.1 fighters, arrived in 1946, and these seventy aircraft were followed in 1949 by sixty SAAB-21R jet fighters, which were jet conversions of the SAAB-21A, a twin-boom fighter with a pusher propeller in the piston form. Other aircraft types followed, contrasting considerably with the neglect of the post-World War I era, and these included two hundred Vampire FB.50 fighter-bombers and T.55 trainers, sixty de Havilland Mosquito NF.19 night-fighters and seventy Supermarine Spitfire PR.19 reconnaissance-fighters. This programme continued throughout the early 1950s, with sixty de Havilland Venom NF.51 jet night-fighters replacing the Mosquito and a large number of SAAB-29 Tunnan ('barrel', an appropriate name!) fighters. SAAB-91 Safir trainers also appeared at this time.

In 1956 the first of 120 Hawker Hunter F.4 fighters appeared, along with the first SAAB-32A Lansen attack aircraft and sixteen Hunting Pembroke C.52 light transport aircraft. The first helicopters, Vertol 44s, arrived in 1957. The SAAB-35 Draken started to replace the Vampires and SAAB-29s in 1959, while currently the SAAB-37 Viggen has been replacing the SAAB-32s, and may replace the Drakens in due course.

Currently the Flygvapnet has some 16,000 men, and is building up to operate ten squadrons of SAAB-37A Viggen attack aircraft, of which there will be 175 initially. Twenty-one interceptor squadrons operate more than three hundred SAAB-35 Drakens; three reconnaissance-fighter squadrons operate Drakens, while another two operate SAAB-32 Lansens; a transport squadron operates two Lockheed C-130E Hercules and seven Douglas C-47 Dakota transports; a helicopter squadron operates ten Boeing-Vertol 107s; while a number of Hawker Hunters and SAAB-29s are employed on target-towing duties; and training is on eighty SAAB-91B Safirs, 130 SAAB-105A and twenty SAAB-105Bs, seventeen SAAB-37 Viggens and a number of SAAB-35 Drakens. Some fifty-eight Scottish Aviation Bulldog trainers are in course of delivery, with the possibility of the order being increased. Six Agusta-Bell 204B and a few Sud Alouette II helicopters are employed on communications duties.

Royal Swedish Navy

The Royal Swedish Navy has been operating Vertol 44K helicopters since 1958, and has overall responsibility for all Swedish military helicopter supply. Currently ten Vertol 107, ten Sud Alouette II and ten Agusta-Bell 205 helicopters are in service. Kawasaki-Vertol KV 107 helicopters may be ordered.

Army Aviation *Armen*

The Swedish Army has had its own air element since 1964 for communications, liaison and AOP duties. Currently, five Dornier Do 27 and twelve Piper L-21 Super Cubs are operated; with nine Sud Alouette II, twelve Agusta-Bell 204B and forty JetRanger, and seven Hughes 269A helicopters. An option is held for forty-five Scottish Aviation Bulldog armed trainers.

SWITZERLAND

Swiss Air Force *Kommando Flieger und Fliegerabwehrtuppen*
Anti-Aircraft Command

Swiss military aviation began in 1914 with the creation of a Fliegertruppe, or Air Troop, with eight aircraft of Blériot,

Morane, Schneider, Aviatik and Henri Farman manufacture. During World War I, in which Switzerland remained neutral, the Swiss-designed Haefeli DH1, DH2 and DH3 observation aircraft were produced. Some Swiss pilots flew with the French Aviation Militaire during the war.

In 1919 the force was reorganized as the Militär-Flugwesen, by which time it had about one hundred aircraft, mainly of Swiss manufacture. Some foreign aircraft were acquired during the 1920s, notably ex-wartime Fokker D.VII fighters, while Swiss aircraft of the period included Haefeli DH5 bombers, followed by the Haefeli M.7 fighters and M.8 bombers. The late 1920s and early 1930s saw licence-built Dewoitine D.9, D.26 and D.27 fighters, Potez 25 general-purpose aircraft and Fokker C.VE reconnaissance-bombers enter service, along with Hawker Hind bombers and de Havilland Moth and Tiger Moth trainers.

Pre-World War II re-equipment included Potez 63 fighter-bombers, ninety Messerschmitt Bf 109E fighters and thirteen Bf 108 liaison aircraft; while Morane-Saulnier M.S.406C fighters and Bücker Bü 131 Jungmann and Bü 133 Jungmeister were licence-built for what had by this time become the Schweizerische Flugwaffe, a separate service. The Swiss Air Force had one hundred fighters and more than one hundred AOP aircraft at the outbreak of the war, during which Switzerland remained neutral, but received some German equipment, including additional Bf 109Es, Fieseler Fi 156C Storch AOP aircraft and Bücker Bü 181 Bestmann trainers.

The end of the war saw one hundred North American F-51D Mustang fighter-bombers enter service, with forty North American T-6 Harvard trainers, and, in 1949 and 1950, seventy-five de Havilland Vampire FB.6 jet fighter-bombers, which were followed by one hundred Swiss-built Vampires to replace the Mustangs. The next aircraft into service was the de Havilland Venom FB.50 fighter-bomber, of which 250 were built in Switzerland for the Swiss Air Force. In 1958, one hundred Hawker Hunter F.58 fighters were placed in service. During the late 1960s, fifty-seven licence-built Dassault Mirage IIIS fighters joined the KFuF (as it is now called) inventory. A number of nationally-designed aircraft entered service during the post-war period on second-line duties, including Pilatus P-2 and P-3 trainers.

Currently the KFuF has some 8,000 men, plus more than 40,000 reservists, and operates two squadrons of Dassault Mirage IIIS interceptors and one squadron of Mirage IIISR reconnaissance-fighters; five squadrons with Hawker Hunter F.58 fighters; thirteen squadrons of de Havilland Venom FB.50 fighter-bombers, some of which are being replaced by second-hand Hawker Hunters pending a decision on a longer-term replacement aircraft; thirty Sud Alouette II and ninety Alouette III and twenty or so Bell 47G helicopters; Bücker Bü 131 and Bü 133, Pilatus P-2 and P-3, North American T-6 Harvard, Beech Twin Bonanza and de Havilland Vampire T.55 trainers; and three Junkers Ju 52/3M, seven Dornier Do 27 and a few other communications aircraft.

SYRIA

Syrian Air Force

Syria gained independence in 1943 having been a French mandated territory, and the Syrian Air Force dates from the final withdrawal of foreign forces in 1946. The first aircraft did not arrive until 1949, and so the Arab-Israeli conflict of the late 1940s by-passed Syria. Initially former Armée de l'Air bases were used, and equipment included Fiat G.46 and G.59, and de Havilland Canada Chipmunk trainers, with French-built Junkers Ju 52/3M, Douglas C-47 Dakota and Beech C-45 transports. A British embargo on arms for the Middle East from 1951 to 1953 did not prevent Syria from obtaining thirty de Havilland Vampire FB.52 jet fighter-bombers via Italy, but these were immediately passed to Egypt. Syria's own first combat aircraft arrived in 1953, and consisted of twenty-three Gloster Meteor F.8 fighters, NF.13 night-fighters, and T.7 trainers, and forty Supermarine Spitfire 22 fighters.

In 1955 the Soviet Union agreed to supply twenty-five Mikoyan MiG-15 fighters and second Soviet personnel for training purposes. However, not only did most of these aircraft never arrive, but the few which did were destroyed on the ground during the 1956 Suez Crisis. Soviet influence in the area was growing, nevertheless, and Syria was able to obtain sixty MiG-17 fighters. Egypt and Syria joined together to form the United Arab Republic in 1958, but Syria withdrew after a *coup d'état*. A number of MiG-15s entered service in the meantime. During the 1960s the Syrian Air Force took delivery of Mikoyan MiG-21 interceptors; Ilyushin Il-14 transports; Mil Mi-1 and Mi-4 helicopters; MiG-15UTI, Yakovlev Yak-11 and Yak-18 trainers; plus Guideline surface-to-air missiles. The June 1967 Arab-Israeli War saw much of this equipment destroyed – probably about 75 per cent of it.

Since 1967, Russia has endeavoured to make good Syria's losses, with the result that the Syrian Air Force is probably now stronger than it ever was.

Currently the Syrian Air Force has at least 10,000 men, and operates ninety Mikoyan MiG-21 interceptors; eighty MiG-15 and MiG-17 and forty Sukhoi Su-7B fighter-bombers; eight Ilyushin Il-14, six Douglas C-47 Dakota and three Lisunov (C-47) Li-2 transports; four Mil Mi-1, eight Mi-4 and four Mi-8 helicopters; with Yakovlev Yak-11 and Yak-18 and MiG-15UTI trainers.

TANZANIA

Tanzanian People's Defence Force Air Wing

Tanzania was derived from the federation of Tanganyika and Zanzibar, two former British colonies, in 1964. The Tanzanian's People's Defence Force Air Wing was formed with Luftwaffe assistance, including an offer of six Nord Noratlas transports, eight Dornier Do 28 liaison and communications aircraft, and nine Piaggio P.149 trainers. This aid ended before deliveries could be completed in 1965 after Tanzania recognized East Germany and East German aid took over. Currently the 250-strong TPDFAW operates an Antonov An-2, five de Havilland Canada DHC-3 Otter and four DHC-4 Caribou transports, and seven Piaggio P.149 trainers; and is expected to receive some Mikoyan MiG-17 fighters from the Soviet Union in the near future.

THAILAND

Royal Thai Air Force

Thailand is situated in a potential cold-war 'hot spot', maintaining a small, but efficient, air force with one of the longer histories of military aviation in the Far East, dating from 1911. The Royal Thai Air Force dates from 1937, when it was formed as the Royal Siamese Air Force, changing its name when the country did so two years later.

Siamese army officers were sent to France for flying training in 1911, returning home in 1913 with four Nieuport and four Blériot aircraft. The Kingdom of Siam entered World War I on the side of the Allies, and sent a contingent of the Siamese Flying Corps with an expeditionary force to Europe. It was not until the end of the war that the first Siamese pilots flew operational sorties, but more than one hundred Siamese officers and NCOs received flying training in France during the war. The Siamese Flying Corps left Europe in 1919, after a spell of serving as part of the Allied Army of Occupation in the Rhineland, with a number of ex-wartime aircraft including Spad S.VII and S.XIII and Nieuport-Delage N.D. 29 fighters, Breguet Br.14A2 and 14B2 reconnaissance-bombers, and Nieuport trainers. In that same year the force's title was changed to the Royal Siamese Aeronautical Service.

Starting in 1920, a domestic airline was operated using converted Breguet Br.14s, which also operated on reconnaissance, survey and liaison duties with the 2nd Group of the RSAS. The 1st (Pursuit) Group concentrated on air defence, using the Spad and Nieuport fighters. Pilots were trained in the United States, United Kingdom, France, Italy and Germany, as well as in Siam. It was not until 1930 that the first post-war aircraft were bought, twenty Avro 504 trainers, while a further fifty of this type were built under licence. During 1930 and 1931, a number of aircraft were evaluated, but no orders placed until 1934, when seventy-two Vought V.100 Corsair AOP aircraft were built in Siam to replace the Breguet Br.14s, followed by twelve each of Curtiss Hawk II and III fighters, while a further twenty-five Hawk IIIs were licence-built.

Further re-equipment took place towards the end of the decade. In 1937 six Martin 139 bombers entered Royal Siamese Air Force service; and in 1939 twenty-five Hawk 75N fighters and some North American NA-69 bombers were ordered, with six North American NA-68 fighters in 1940; but only the Hawks were delivered since the NA-69s and NA-68s were taken off their ships at the Philippines and Hawaii respectively to be pressed into US service. Some support in Thailand for Japan led to the procurement of nine Mitsubishi Ki 21 bombers and nine Tachikawa Ki 55 trainers.

An invasion of French Indo-China was launched by Thailand in January 1941, after a border dispute, and this brought French and Thai air elements into combat until Japan arranged a truce in May, followed by the Vichy French Government allowing Japanese use of bases in French Indo-China. Japan invaded Thailand in December and, although the Thai forces fought valiantly against overwhelming odds, defeat was inevitable and the Thai Government arranged a cease-fire and surrender within a few days.

During the rest of World War II, Thailand was officially an ally of Japan, but Royal Thai Air Force personnel were restricted to a non-combatant role. The more reliable members of the RTAF helped the underground movement, which was controlled from RTAF headquarters, flew Allied agents into, and Allied pilots out of, Thailand, and avoided suspicion even though Allied bombers took care to minimize damage to Thai equipment and installations when attacking Japanese forces. Pro-Allied and pro-Japanese personnel were strictly segregated, the latter being sent with Japanese equipment to the north of the country which was under heavy Allied air attack.

Towards the end of the war a few combat aircraft were supplied to the RTAF, including about a dozen each of Nakajima Ki 27 and Ki 43 fighters, nine Mitsubishi Ki 30 bombers and some Mansyu Ki 79 trainers. The return of peace found the RTAF operating some abandoned Japanese aircraft, plus a few of its pre-war Hawks. RAF personnel were seconded to assist the RTAF, and thirty Supermarine Spitfire Mk.14s were placed in service with some Fairey Firefly FR.13s for naval flying, followed by twenty Miles Magisters, forty-two North American T-6G

Texans, some de Havilland Tiger Moths and a number of de Havilland Canada Chipmunks for training.

Thailand was a founder-member of the South-East Asia Treaty Organization in 1954, but even before this the United States had started to provide Thailand with military aid. This included the training of some RTAF personnel in the United States, and the supply of aircraft, including an initial batch of fifty Grumman F8F-1 and F8F-1B Bearcat fighter-bombers and an additional sixty T-6G Texan trainers; Stinson L-5 Sentinels, Piper L-18 Super Cubs and Cessna O-1 Bird Dogs for AOP duties; Fairchild 24Ws, Cessna 170s, and Beech C45s for communications; Westland S.51 Dragonfly, Sikorsky S-55 and Hiller 360 helicopters; and in 1957 a further seventy-five Texans, the first jet aircraft, thirty Republic F-84G Thunderjet fighter-bombers and Lockheed T-33A trainers. The Thunderjets and Bearcats were replaced in 1962 by sixty North American F-86F Sabres, with additional F-86C Sabres in 1966 along with the first of the RTAF's Northrop F-5 Freedom Fighters. Other aircraft delivered during the period included Fairchild C-123 Provider transports, Sikorsky CH-34C helicopters and Cessna C-37B trainers.

Currently the RTAF has some 25,000 men, and twenty North American F-86C and forty F-86F Sabre, and twenty-five Northrop F-5A Freedom Fighter fighter-bombers; five Lockheed RT-33A tactical reconnaissance aircraft; forty North American T-28D Trojan and twenty T-6G Texan armed-trainers, making thirteen combat squadrons in all; five Beech C-45, twenty Douglas C-47 Dakota and two C-54 Skymaster and six Fairchild C-123 transports; fifty Bell UH-1H Iroquois, twenty-two Sikorsky CH-34C and thirteen UH-19, three Kaman HH-43B Huskie and sixteen Kawasaki KH-4 helicopters; with Cessna T-37B, Lockheed T-33A and de Havilland Canada Chipmunk and North American T.6G Texan trainers. Some Grumman HU-16 Albatross amphibians are operated.

Royal Thai Navy

The Royal Thai Navy operates one squadron with Grumman S-2F Trackers and HU-16 Albatross amphibians on maritime-reconnaissance duties.

Royal Thai Border Police

This is a para-military force with some 7,000 men; and ten Bell 204B and eleven 205, ten Hiller UH-12B, a Sikorsky S-55 and a S-62A helicopters are operated on support duties.

TOGO

Togo Air Force *Force Aérienne Togolaise*

The former French colony of Togo was granted independence in 1960, although remaining outside of the French Community apart from a co-operative agreement dating from 1963. The standard arms 'package' of a Douglas C-47 Dakota transport and two Max Holste 1521M Broussards is operated.

TUNISIA

Tunisian Air Force

Tunisia became independent in 1956, and soon afterwards the Tunisian Air Force was formed with Swedish help, although the first aircraft, fifteen SAAB-19D Safir trainers, did not arrive until 1960. Two Alouette II helicopters were placed in service in 1962, and in 1966 eight Aermacchi MB.326B jet trainers arrived from Italy. These have been followed by North American F-86F Sabre fighters.

Currently the Tunisian Air Force has some six hundred men, and operates twelve North American F-86F Sabre fighters, eight Aermacchi MB.326B armed jet trainers, twelve North American T-6G Texan and fourteen SAAB-91D trainers, supported by three Dassault Flamant transports and eight Sud Alouette II helicopters.

TURKEY

Turkish Air Force *Türk Hava Kuvvetleri*

Turkey has the longest history of military aviation of any Middle Eastern nation, dating from 1912 when the Government ordered an assortment of Bristols, D.F.W.s, Mars, Déperdussins, Nieuports and R.E.P.s for army co-operation duties, on which these were flown by foreign pilots. In 1914 this service became the Turkish Flying Corps as the result of a German initiative and Turkey fought on the side of the Central Powers against the Allies during World War I. Aircraft flown by the Turkish Flying Corps during the war included A.E.G. C.IV and Albatros reconnaissance aircraft, Halberstadt D.II fighters and some Gotha WD.13 seaplanes for navy flying. All of these aircraft were flown by German pilots with Turkish observers and navigators.

The defeat of Germany and her allies in 1918 meant that, under the Treaty of Versailles, all military aviation was forbidden to the former Central Powers. However, in 1925 the Türk Hava Kurumu (Turkish Air League) was formed with two aircraft, a Caudron training biplane and an Ansaldo A.300, purchased by public subscription. The THK was mainly French-assisted during its early years, although some training did take place in the United Kingdom. Morane-Saulnier M.S.53 trainers were delivered in late 1926, and in 1928 a small number of Ruhrbach Ru III flying-boats were delivered, followed by eighteen Curtiss Hawk fighters and some Fledgling trainers, with subsequent licence-production.

Development of the THK really got under way during the 1930s with the delivery of twenty Breguet Br.19B.2 reconnaissance-bombers and six Supermarine Southampton maritime-reconnaissance aircraft. In 1937 orders were placed for thirty each of Heinkel He 111D, Bristol Blenheim I and Martin 139 bombers, as well as for Supermarine Walrus amphibians, Avro Anson bombers and Vultee V-IIG fighter-bombers; Hanriot 182 and Westland Lysander army co-operation aircraft; Miles Hawk II and Curtiss-Wright C.W.22 trainers, and Focke-Wulfe Fw 58 Weihe communications aircraft. Hawker Hurricanes and additional Blenheims were ordered in 1938. Other aircraft of this period included de Havilland Dragon and Dragon Rapide light transport aircraft, which were also used on photographic and navigational training duties. Licence-built Gotha trainers were also in service, but an order for forty Gotha G-23 fighters was never fulfilled – the aircraft were delivered to the Spanish Republicans.

Turkey remained neutral during World War II, signing a non-aggression pact with Germany and receiving aircraft from both sides, including additional supplies of those already in service, and also Curtiss Tomahawk IIBs, Fairey Battles, Airspeed Oxfords, Morane-Saulnier M.S.406 and Focke-Wulfe Fw 190As. The RAF also supplied spares for Heinkel He 111s salvaged from aircraft shot down over the United Kingdom. Lend-lease equipment was also supplied, including Supermarine Spitfire VBs, Hawker Hurricanes, Bristol Beaufighter Is, Beaufort Is and Blenheims, Martin Baltimores and Consolidated Liberators.

After the cessation of hostilities, the THK received a large number of war-surplus aircraft, many of which were of the same types as those introduced before and during the war, but also with some de Havilland Mosquito T.3 trainers and FB.6 fighter-bombers, Republic F-47D Thunderbolt fighter-bombers, North American T-6 Harvard and Beech T-11B Kansan trainers, Douglas B-26 Invader bombers, and C-47 Dakota transports, with a few Beech D.18 transports. A large proportion of these deliveries were covered by the United States Military Aid Programme after Turkey became a member of the North Atlantic Treaty Organization in 1952. USAF personnel were also seconded as advisers and instructors.

Turkey received her first jet aircraft in 1952 with the arrival of the first of three hundred Republic F-84G Thunderjet fighter-bombers, accompanied by twenty-four Beech T-34 Mentor trainers, the latter being built in Turkey. In 1953, Canadair F-86E Sabre Mk. 2 and 4 fighters were obtained, at the same time as the first of Turkey's own M.K.E.K. Ugar (Lark) trainers. Since this period, deliveries of American aircraft have continued, including Lockheed F-104G Starfighter interceptors, RT-33A tactical reconnaissance aircraft and T-33A trainers and C-130E

Hercules transports; Cessna T-37 trainers; Dornier Do 27 and Do 28 communications aircraft, Piper L-18 AOP aircraft, and Northrop F-5A tactical fighter-bombers.

Currently the THK has some 50,000 men, and operates one squadron of twenty-five Convair F-102A Delta Dagger and two squadrons with a total of thirty-six Lockheed F-104G interceptors; six squadrons with 140 Northrop F-5A Freedom Fighter tactical fighter-bombers; six fighter squadrons with 125 North American F-86D/E/K Sabres; ten North American F-100C Super Sabre fighter-bomber squadrons with two hundred aircraft; three Republic RF-84F Thunderflash reconnaissance squadrons; four transport squadrons with six Beech C-45, ten Douglas C-47 Dakota and three C-54 Skymaster, and ten Lockheed C-130E aircraft; a small number of Dornier Do 27 and Do 28 communications aircraft, in which role also operate some Agusta-Bell 204B and JetRanger helicopters; while for training there are Beech T.34 Mentors and T-11 Kansans, forty North American T-6 Harvards, thirty Lockheed T-33As, and twenty-three Cessna T-37s. Two battalions of Nike-Hercules surface-to-air missiles are deployed.

Turkish Army

The Turkish Army operates a number of Agusta-Bell 47G Sioux and 204B helicopters.

UGANDA

Uganda Army Air Force/Police Air Wing

Uganda was granted independence by Great Britain in 1962, and in 1964 an Army Air Wing was formed to supplement the work of the Police Air Wing, which had been formed before independence and was largely operated by European personnel at that time. Initially the Police Air Wing Westland Scout helicopters were shared, while aid was received from Israel, but Russian and Czech aid took over during the latter half of the 1960s.

Currently the 450-strong UAAF/PAW operates seven Mikoyan

MiG-15 fighter-bombers; twelve Potez Magister armed-trainers; six Douglas C-47 Dakota and a de Havilland Canada DHC-4 Caribou transport aircraft; five Piper L-18 Super Cub AOP aircraft; an Aztec and four Cessna 180 communications aircraft; four Piaggio P.149 and five L-29 Delfin trainers; while there are also two Westland Scout helicopters which are used mainly on police work.

UNION OF SOVIET SOCIALIST REPUBLICS

Soviet Military Aviation Forces *Sovietskaya Voenno-Vozdushnye Sily*

Russian military aviation dates from the formation of a Central Aviation School in 1910, although the Army was already using observation balloons by this time. In 1912 the Imperial Government bought small numbers of British and French aircraft for the Army, and licensed production of these types resulted in some 240 aircraft of Bristol, Farman, Morane and Nieuport design being in Imperial Russian Flying Corps service at the time World War I started in 1914. The following year saw the first Russian-designed fighters, Sikorsky's RBVZ-S-16, RBVZ-S-17 and RBVZ-S-20, for the IRFC. Other types were not far behind, including Lebed-7 and Lebed-10 fighters, Lebed-12, Anasal and Anade reconnaissance aircraft, and Sikorsky's Ilya Mourometz bomber. Nevertheless, foreign designs predominated in IRFC service, including Sopwith 1½–Strutter, B.E.2e and R.E.8 fighters, D.H.4 and D.H.9A bombers. The IRFC's war differed from that of the Western Front air forces in being mainly reconnaissance and bombing, with little fighter activity.

Probably the most loyal of the armed forces, the IRFC did not pass into the hands of the Bolshevik revolutionaries until late in 1917. In 1918 the Glavnoe Upravlenie Raboche-Krestyanskogo Krasnogo Vozdushnogo Flota – GUR-KKVF – (Workers and Peasants Red Air Fleet) was formed; it was known as the Red Air Fleet until the Sovietskaya Voenno-Vozdushnye Sily (Soviet Military Aviation Forces) was formed in 1924. At the end of the revolution some three hundred aircraft were in Red Air Fleet service, and during the next few years large numbers of foreign aircraft were ordered, including Fokker D.XIII fighters and C.IV reconnaissance and some Ansaldo types. Also D.H.9A bombers were produced in Russia as the R.1 and several German designs also entered production since Germany was not allowed to produce military aircraft under the terms of the Treaty of Versailles. Imperial Russia's most notable aircraft designer, Sikorsky, had been forced by the Russian Revolution to leave for the United States.

The late 1920s saw the first post-revolution Soviet designs, the Polikarpov I-1 and I-2 and the Grigorovich I-1 and I-2 fighters, the Tupolev R-3 and R-6 reconnaissance aircraft, TB-1 bomber and G-1 transport. The next few years saw a succession of Russian designs enter production and service, the Polikarpov I-3 and Tupolev I-4, Polikarpov-Grigorovich I-5, Grigorovich I-6 and DI-3 fighters; Polikarpov TB-2 bomber; R-5 reconnaissance and V-2 training aircraft; and Tupolev G-2 transport. by 1930 there were 1,000 aircraft in twenty regiments, and the SV-VS had had its first taste of combat in a limited war on the Chinese border in 1929.

The 1930s were marked by large orders for heavy bombers, notably the Tupolev TB-2 and the Kalinin K-7, while Tupolev also designed the SCh-1 and SCh-2 ground-attack aircraft, the SB-2 light bomber and the I-14 fighter. Other aircraft of the period included the Ilyushin TB-3 bomber, the Polikarpov I-15, I-15B and I-16 fighters. It must be noted that, as in Russia since the revolution the state owns everything, the names attached to an aircraft are those of the designers or the heads of design teams, rather than those of the manufacturers as elsewhere. In 1936, SV-VS personnel and aircraft were sent to Spain to fight on the Republican side in the Spanish Civil War, during which a total of 1,400 of the latest Soviet fighter and bomber aircraft were sent to Spain, and their performance has since been claimed to be as good as that of their German and Italian opposition, although the Republicans lost the war!

The next opportunity for aerial combat came against the Japanese in 1938 and 1939, and the same aircraft types which had seen service in Spain were used again, and although Japanese forces retreated from Mongolia in 1939, it was at the expense of heavy Soviet casualties. From 1937 onwards, large numbers of Russian aircraft were provided for Chinese Communist forces, often with 'volunteer' pilots and ground crew. The SV-VS during the late 1930s was a force with some 6,000 aircraft of all types; but efficiency was frequently compromised by the 'purging' of officers on political grounds, usually with subsequent imprisonment and execution for those concerned. A few American and German aircraft were also obtained during this period to give Soviet aircraft designers a desperately needed insight into Western methods, and in 1938 a Russo-German agreement resulted in further supplies of Messerschmitt, Heinkel and Dornier types. The Polikarpov I-17 fighter entered service before the start of World War II in only limited numbers, as were examples of ground-attack and dive-bombing types, including the Polikarpov VIT-1 and VIT-2, the Ilyushin Il-2, the Arkhangelskii Ar-2, and the Tupolev R-10 reconnaissance-bomber.

In 1939, Soviet forces invaded Finland, but the Russo-Finnish War ended the following year after 2,000 Russian aircraft met determined Finnish and Swedish resistance. Never short of a variety of new aircraft types at any one time, the period at the start of World War II was no exception, and there were Lavochkin I-22, Yakovlev I-26 (Yak-1 when the designations were changed shortly afterwards to that still current, reflecting designs rather than types), Yatsenko I-28, Mikoyan I-61 and I-200 (later respectively MiG-1 and MiG-3) fighters; the Sukhoi Su-1 strike aircraft; the Ilyushin DB-3 (Il-4), Petlyakov Pe-2 and Pe-8, and Yermolaev Yer-2 and Yer-6 bombers; the Antonov A-7 transport and SS-2 liaison aircraft. Although, at the start of the war, Germany and Russia had been allies, Russia changed sides and, in June 1941, Germany attacked Russia, with some 2,500 Luftwaffe aircraft pitted against some 10,000 SV-VS aircraft. Then, as now, the SV-VS was the largest and most significant part of the total Soviet air power.

Only Allied military aid saved Russia from a successful German invasion. The numerical superiority of the SV-VS was no match for the technical superiority of the Luftwaffe, and it was not until large numbers of British and American aircraft were actually in front-line service that the Luftwaffe could be effectively countered. From the United Kingdom the SV-VS received

Hawker Hurricane and Supermarine Spitfire fighters, de Havilland Mosquito fighter-bombers, and Armstrong-Whitworth Albermarle AOP aircraft. From the United States came Bell P-39N Airacobra and P-63C Kingcobra, Curtiss P-40 and Republic P-47 Thunderbolt fighter-bombers; Douglas A-20 ground-attack aircraft; North American B-25 Mitchell bombers and AT-6 trainers; Consolidated PBY-5A Catalina amphibians; and Douglas C-47 Dakota transports, which were also built in Russia as the Lisunov Li-2. In all, some 15,000 aircraft were given to the USSR by the two Allied powers. The result was that Russia was able to go over to the offensive and carry the attack through to Berlin in 1945.

Towards the end of the war, new and improved Russian types entered service; these included Yakovlev Yak-3 and Yak-9, Lavochkin La-7 and La-9 and Mikoyan MiG-5 fighters, Tupolev Tu-2 bombers (supposedly designed by Tupolev during a period of imprisonment under the Stalin regime) and Ilyushin Il-10 ground-attack aircraft. The SV-VS had 20,000 aircraft when World War II ended.

Captured German airframes and engines were shipped to Russia, and the Yakovlev Yak-15 and Mikoyan MiG-9 jet fighters were designed in haste, and followed by the Yakovlev Yak-17. But it was not until a number of Rolls-Royce Nene and Derwent engines were purchased from Great Britain, for copying by the Russian aircraft industry, that the famous Mikoyan MiG-15 fighter could be designed, based on captured German airframe plans. In the meantime, Boeing B-29 Superfortresses, which had force-landed in Russia, were copied and put into production as the Tupolev Tu-4. Other less well-known aircraft of this period included the Yakovlev Yak-23 and Lavochkin La-15 jet fighters, and the Tupolev Tu-77 twin-jet attack aircraft, from which the Tu-12 and Tu-14, and eventually the Ilyushin Il-28 jet bombers were developed. The famous Antonov An-2 utility transport first flew in 1947.

Once again a convenient opportunity for the Soviet Union to test its new aircraft occurred: this time in Korea in 1950, when the Soviet Union fought against United Nations forces comprising British, American, Australian and South African contingents. The Mikoyan MiG-15 was used against British Gloster Meteor and United States North American F-86 Sabre jets, as well as Fairey Fireflies and North American F-51D Mustangs, both types of which accounted for at least one MiG-15 each! The MiG-15's deficiencies led to the MiG-17 in 1952, at the same time as a MiG-15UTI conversion trainer. During the early 1950s, deliveries of aircraft began to the 'satellite' countries invaded by Russia towards the end of World War II, or allied with Russia after a *coup d'état*, with Poland and China being allowed to assemble Russian types. A Sukhoi experimental aircraft broke the sound barrier, the first Russian aircraft to do so, and in 1955 the supersonic Mikoyan MiG-19 fighter appeared. Other aircraft of this period included the Yakovlev Yak-25 night-fighter; the Tupolev Tu-16 long-range turboprop bomber and the Tu-20 jet bomber; Ilyushin Il-12 and Il-14 transports; the Mil Mi-1 helicopter of 1950, followed by the Mi-4 in 1951, the Yak-4 in 1954, the Mi-6 in 1958, and Mi-8 and Mi-10 helicopters later. Amongst the bombers can also be counted the large Myasishchev Mya-4, introduced in 1955, while transports included the Antonov An-12. The Mikoyan MiG-21 interceptor started to enter service in 1958, at the same time as the Sukhoi Su-7 ground-attack aircraft, and both types are standard equipment for Eastern bloc air forces. Indeed most of the aircraft of the period are those of today. The Sukhoi Su-9 is a development of the Su-7 for interceptor duties.

The most recent developments have been the entry into service, from 1968 onwards of the Mikoyan MiG-23 Mach 3 interceptor and, during the mid-1960s, of Tupolev Tu-22 rear-engined bombers. All Soviet aircraft are given NATO code names, and these are mentioned in the aircraft section of this book.

Currently the SV-VS has some 500,000 men, and operates Mikoyan MiG-23 and MiG-21 interceptors, with a few MiG-19 and MiG-17 fighters now deployed in the tactical role with Sukhoi Su-7 and Yakovlev Yak-28 ground-attack and light bomber types respectively; Sukhoi Su-9 all-weather interceptors are also in service, possibly with a few Yakovlev Yak-25s remaining in this role; there are about ninety Myasishchev Mya-4 and one hundred Tupolev Tu-20 long-range bombers and tankers; about six hundred Tupolev Tu-16 and two hundred Tu-22 medium-range bombers, with possibly some Ilyushin Il-28s surviving in this role. A total of just under 2,000 transports excludes the airline Aeroflot, which is organized on military lines, and the types used include Ilyushin Il-14 and Il-18, Tupolev Tu-104, Tu-114, Tu-124 and Tu-134, Antonov An-2, An-12, and An-22. There are about six hundred helicopters in addition, including Mil Mi-4, Mi-6, Mi-8, Mi-10 and Mi-12; and training is on Yakovlev Yak-11, Mikoyan MiG-15UTI and MiG-21UTI, Sukhoi Su-7UTI, Ilyushin Il-28U and Il-14U, and L-29 Delfin aircraft.

National Air Defence *Protivo-Vozdushniya Oborona Strany*

The P-VOS developed out of the SV-VS during the late 1930s, and was charged with the duty of protecting the USSR from bomber attack and reconnaissance. Undoubtedly, many of the fighter aircraft supplied to the USSR during World War II will have been placed in P-VOS service, while it will also have received its share of the post-war fighter and interceptor types. Unlike the SV-VS, the P-VOS is unlikely to be deployed outside of the Soviet Union. In 1960 its strength was almost doubled to 4,000 aircraft by the addition of the fighter and interceptor aircraft and personnel of the Soviet Navy.

Currently it is believed that the P-VOS has some 500,000 men and 3,300 aircraft, there having been a reduction in recent years due to the increasing use of surface-to-air missiles. Indeed, between 200,000 and 300,000 of the total available manpower is deployed on anti-aircraft guns and missiles, including the SA-2 Guideline, SA-3 Goa, SA-4 Ganef, SA-5 Griffon and SA-6 Gainful (Goa being short-range, Ganef mobile, Griffon long-range, while Gainful is basically a Ganef replacement). Aircraft include the new Mikoyan MiG-23 and the Sukhoi Su-11 interceptors, plus the older MiG-21 and Sukhoi Su-9 which still form the backbone of the force; a few MiG-19 and MiG-17 fighters are still operational, as are a few Yakovlev Yak-28P and Tupolev Tu-28 all-weather fighters, while Tupolev Tu-114 airborne-early-warning aircraft are also in service. Training is on MiG-15UTI and MiG-21UTI types, with some Yakovlev Yak-11s and L-29 Delfins for basic and basic-jet training respectively.

Independent Naval Air Fleet *Aviatsiya-Voenno Morkskikh Flota*

Soviet Naval Aviation is as old as the country's history of military aviation generally, dating from 1910. In 1912, Curtiss flying-boats were obtained, and by the outbreak of World War I some sixty flying-boats were in service. Grigorovich M-5 and M-9 flying-boats were placed in service during the war, but units of the Imperial Russian Navy were amongst the first to join the Russian Revolution in 1917.

During the early 1920s, Tupolev TB-1P seaplane bombers were supplied to the Soviet Navy, followed by Tupolev MDR-2 long-range flying-boats and Beriev MBR-2 short-range flying-boats and, shortly before the start of World War II, Tupolev MK-1 six-engined flying-boats. Also during this period, the Navy's air arm ceased to be a part of the SV-VS and adopted its present title and structure. Fighters and bombers were operated prior to 1960, when the fighters were transferred to the P-VOS.

The Soviet Navy has never operated aircraft carriers, but two 18,000-ton helicopter cruisers are now in service, and each can take twenty Kamov Ka-25A, Hormone, helicopters.

Currently the A-VMF has some 75,000 men, and operates two hundred Tupolev Tu-16, fifty Tu-20 and one hundred Tu-22 bombers on long-range maritime-reconnaissance duties; fifty Ilyushin Il-28 medium-range bombers; sixty each of Beriev Be-6 flying-boats and Beriev Be-12 amphibians; possibly several hundred Mil Mi-4 and Kamov Ka-25A helicopters; with a number of transport and communications aircraft.

UNITED KINGDOM OF GREAT BRITAIN AND NORTHERN IRELAND

Royal Air Force

Britain has the longest continuous history of military aviation of any nation, dating from 1878 when the Royal Engineers began experimenting with balloons at Woolwich Arsenal, and used these in expeditions to Bechuanaland in 1884 and to the Sudan in the following year. The first real military use occurred in the

South African War of 1889 when balloons were used for observation and artillery control duties. In 1890 sheds were built at Farnborough, where in 1907 both the first British airship and the first British aeroplane were built. The next step was the upgrading of the balloon section to the status of Air Battalion in 1911, but in the meantime the Royal Navy had started flying in 1909.

The Royal Flying Corps, the direct predecessor of the Royal Air Force, was formed in 1912 with the amalgamation of the Air Battalion, Royal Engineers, and the Royal Navy's Air Branch, although the latter left the RFC in 1914 to re-establish itself as the Royal Naval Air Service.

At the outbreak of World War I in 1914, the RNAS had one hundred aircraft, plus a number of airships which were useful as convoy escorts, and the RFC had 180 aircraft. At that time it was envisaged that the RFC would fly reconnaissance missions while the RNAS would defend Britain from aerial attack, even though only a handful of its aircraft were equipped with anything heavier than a rifle or a revolver! A good selection of British and French aircraft types were in service at the outbreak of war, including Avro, Bristol, Short, Sopwith, Royal Aircraft Factory, Blériot, Déperdussin and Farman products.

During the war, RNAS aircraft bombed Zeppelin sheds at Hamburg, Cologne and Friedrichshafen, and torpedoed a Turkish warship in the Mediterranean, as some of its most prominent encounters. Fighter warfare evolved during 1915, and in 1916 German Fokker aircraft, equipped with synchronized Lewis guns (which could fire through the propeller disc without the need for deflector plates fitted to the propeller) gained air supremacy until challenged by the pusher-propeller Vickers F.E.2B and D.H.2 aircraft, and later by the Lewis-gun-fitted French Nieuport biplanes and British Bristol Scouts and Sopwith 1½-Strutters with synchronized Vickers machine-guns. During the latter part of the war the bomber began to evolve as a distinct aircraft type with the appearance of the D.H.9, D.H.9B, Handley Page O/400 and V/1500, and, at the end of the war, Vickers Vimy (first aircraft to fly across the Atlantic in 1919). During the war, bomb size grew from 112lb. to 1,650lb.

The RNAS and the RFC amalgamated on 1 April 1918 to

form the Royal Air Force, the first truly separate air arm in the world. In November, at the end of the war, the RAF had some 360,000 men, 200 squadrons and 23,000 aircraft. However, postwar arms cuts, and the need to submerge the old inter-service rivalries by starting from scratch, led to a reduction to a mere twelve squadrons; one in Germany, two in Britain, and nine in the Middle East and India. This was to be increased to fifteen fighter and thirty-seven bomber squadrons, but this strength was not achieved for some time.

In 1923 the Fairey Fox, the first post-war fighter, joined the ex-war assortment of Sopwith Snipes, D.H.9s, Vickers Vimys and Bristol F.2 fighters, some of which survived until 1929 when the front-line aircraft were replaced by Gloster Glebes, Armstrong-Whitworth Siskins and Vickers Virginias. By 1933 there were seventy-four regular squadrons and thirteen auxiliary squadrons. Other aircraft of the period included Boulton Paul Sidestrands, Fairey IIKs, Handley Page Hinaidis and Hyderabads, Hawker Harts and Horsleys, Blackburn Iris and Supermarine Southamptons. All of these aircraft were biplanes, as was the Hawker Fury with a maximum speed of 225 mph, which was introduced to service in 1931 and was the RAF's front-line fighter for the next few years.

The inter-war period saw the RAF engaged in a number of police actions abroad, and operating a number of mail services prior to the formation of Imperial Airways, the first of these flying from Baghdad to Cairo in 1921. In 1924 the aircraft carrier-borne detachments of the RAF became the Fleet Air Arm.

In 1936 and 1937, reorganization took place with the division of the RAF into Bomber, Coastal, Fighter, Maintenance and Training Commands in 1936, and the Admiralty regained full responsibility for all naval aviation in the following year. A plan evolved in 1936 called for the RAF's regular strength to be increased from seventy-four to 134 squadrons, and the auxiliary squadrons to be increased from fifty-two to 138. Development and production of a number of new aircraft was put in hand, these including the Armstrong-Whitworth Whitley, Fairey Battle, Bristol Blenheim and Vickers Wellington bombers, and the Hawker Hurricane and Supermarine Spitfire fighters. But it was not until 1938 that industrial capacity replaced finance as the limiting factor in the RAF's expansion, so that by 1939 the RAF still had only one-eighth of the manpower and two-sevenths of the equipment of the Luftwaffe.

At the outbreak of World War II, Bomber Command possessed fifty-five squadrons, including five with Armstrong-Whitworth Whitley IIIs and IVs, six with Handley Page Hampdens, six with Bristol Blenheim IVs, six with Vickers Wellington Is, ten with Fairey Battles, while the rest had various obsolescent and obsolete types. Coastal Command possessed ten squadrons of Avro Ansons, one of Lockheed Hudsons, two of Short Sunderland flying-boats, six of obsolete Supermarine Stranraer and Saunders-Roe London biplane flying-boats and Vickers Vildebeeste torpedo-bombers. Fighter Command had twenty-two squadrons of Hawker Hurricanes and Supermarine Spitfires, and thirteen of obsolete Gloster Gladiator biplanes. Some twenty-seven squadrons were based in France, where six hundred French and British fighters had to face 3,000 German aircraft. Another twenty-seven squadrons were based in the Middle East, where in spite of operating mainly obsolete aircraft they managed to gain air supremacy and help defeat the Italian attack on Greece in 1940, though the German attack in the following year was successful.

The RAF was unable to provide more than token support for the retreat from Norway and in 1940 suffered its most testing period during the Battle of Britain, which it won in spite of overwhelming odds. However, after mid-1941 the RAF gained an increasing aerial supremacy which was bolstered by new types of aircraft then entering service, including four-engined heavy bombers, Avro Lancasters, Handley Page Halifaxes and Short Stirlings, and the de Havilland Mosquito fast fighter-bomber, and Bristol Beaufighter long-range fighter. The famous thousand bomber raids on Berlin began at this time. Amongst the most famous actions of the entire war was the raid by 617 Squadron's Avro Lancasters on the Ruhr dams in 1943.

Japan's successes in the Pacific War, however, put the RAF at a severe disadvantage in the Far East, eventually compelling the force to withdraw to India, Australia and New Zealand after the fall of Singapore and Malaya.

Towards the end of the war in Europe, American Liberators, Bostons and Fortresses entered RAF service, supplementing Bomber Command's British aircraft. In March 1943, Transport Command was formed. One interesting RAF unit formed during the mid-war years was the Merchant Service Fighter Unit, operating Supermarine Spitfire and Hawker Hurricane fighters which were catapulted from merchant vessels and landed in the sea after combat, whence the pilot was picked up by an escort vessel. The RAF also provided tows for the Hamilcar and Horsa gliders which were used during the Normandy invasion in 1944, and raided the launching sites for the V.1 and V.2 rockets. New equipment at the end of the war included Hawker Typhoons and Tempests, Lockheed Lightnings and North American Mustangs, and Avro Lincolns, the last-named being heavy bombers. In 1944 the RAF became the first air force to put jet fighter aircraft into operation, with the arrival of the Gloster Meteor.

At the end of the war in Europe the RAF had some 1,100,000 men; 487 squadrons, of which a hundred came from the Dominions or the 'Free' air forces; and 9,200 aircraft. Heavy post-war commitments prevented the RAF's strength from being immeiately reduced to the peacetime target of 300,000 men. The last piston-engined fighter, the de Havilland Hornet, entered service, and by 1947, virtually all Lancaster bombers had been replaced by Lincolns; while replacement of Douglas C-47 Dakotas and Avro Yorks by Vickers Valettas and Handley Page Hastings respectively was delayed by the Berlin Airlift in 1948, when Russia tried to isolate West Berlin from West Germany.

Amongst the post-war duties can be included police actions in Palestine and Malaya. In 1950, with a strength of 202,500 men, the RAF received de Havilland Venom fighters and the world's first jet bomber, the English Electric Canberra, while a few Boeing B-29 Superfortressess and Lockheed Neptune maritime-reconnaissance aircraft were used while waiting for the first Avro Shackletons to enter service with Coastal Command. Only limited RAF participation in the Korean War was possible because of the lack of manpower and equipment; but in 1951 came a re-armament programme, and the RAF's strength was increased to 250,000, although Royal Auxiliary Air Force squadrons had to be called up for three months at a time in order to achieve this. The United Kingdom at this time was also a founder-member of the North Atlantic Treaty Organization.

RAF aircraft of the 1950s included the first military jet transport aircraft, the de Havilland Comet C.2, the RAF's first turboprop aircraft, the Bristol Britannia transport, and the world's first turboprop freighter, the Armstrong-Whitworth Argosy, while Blackburn Beverley tactical transports also entered service. Westland Whirlwind helicopters were introduced, to be followed almost a decade later by a few Westland Wessex helicopters, while the heavier Bristol Belvedere served throughout the late 1950s and 1960s. The RAF's first swept-wing fighter, the Supermarine Swift, entered service in 1954, and was withdrawn because of technical difficulties after only a few months, but the Hawker Hunter which followed shortly afterwards proved to be a tremendous success. Gloster Javelin delta-wing interceptors entered service during the mid-1950s, which period also saw de Havilland Canada Chipmunks and Percival Provost trainers in service, while Jet Provosts replaced Vampire jet trainers, and Folland Gnat advanced jet trainers replaced Meteors. The famous 'V' bombers, the trio of Vickers Valiant, Avro Vulcan and Handley Page Victor long-range four-engined jet nuclear bombers also became operational, to carry Britain's nuclear deterrent. Initially this consisted of atomic and later hydrogen bombs, but stand-off missiles were later used.

The main action for the RAF during the decade was in the Suez Canal zone with French and Israeli forces in 1956. During the early 1960s, the RAF supported ground forces fighting Indonesian infiltrators attempting to sabotage the newly-created Federation of Malaysia.

This should not be allowed to conceal the fact that the late 1950s and early 1960s were a period of drastic reduction in RAF strength. New aircraft continued to enter service, notably the English Electric Lightning interceptor, Short Belfast, Vickers VC 10 and Hawker Siddeley Andover transports, followed during the mid-1960s by Lockheed C-130K Hercules transports, McDonnell Douglas F-4M Phantom II fighter-bombers and Hawker Siddeley Harrier vertical take-off strike aircraft. During

1970, Hawker Siddeley Buccaneer low-level strike-bombers entered service, as did the first Whirlwind-replacement Sud-Westland SA.330 Puma helicopters. The early 1970s will see BAC-Breguet Jaguar strike aircraft enter service, with Panavia 200 Panthers following in about 1975 – the RAF is due to receive some four hundred of this multi-role combat swing-wing aircraft.

Currently the RAF has some 110,000 men and women in four commands: Strike Command (formerly Fighter, Bomber and Coastal Commands), Air Support Command (formerly Transport Command plus some former Fighter Command units), Training Command and Maintenance Command. RAF aircraft outside of the United Kingdom usually come under the control of Near East Command, Far East Command, RAF Germany and RAF Gulf. There are about one hundred BAC Lightning interceptors; 146 McDonnell Douglas F-4M/K Phantom II fighter-bombers (to which another twenty-four Fleet Air Arm F-4Ks will be added after the disposal of the last aircraft carrier around 1975); seventy-seven Hawker Siddeley Harrier vertical take-off strike aircraft; thirty-five Hawker Siddeley Buccaneer S.2B low-level strike bombers (to which sixty-four FAA aircraft will be added in the mid-1970s); fifty Hawker Siddeley Vulcan nuclear bombers, usually deployed with Blue Steel air-to-surface missiles; twenty-four Handley Page Victor tanker aircraft, with another sixteen in process of conversion from the bomber role; numbers of BAC Canberra light jet bombers and photographic reconnaissance aircraft, which are being replaced by Phantoms and Buccaneers for the most part; while remaining Hawker Hunters await delivery of 164 BAC-Breguet Jaguar strike aircraft during the early 1970s. There are thirty-eight Hawker Siddeley Nimrod maritime-reconnaissance aircraft, with some remaining Avro Shackletons on this duty and also airborne early warning; ten Short Belfast, fourteen Vickers VC 10, sixty-six Lockheed C-130K Hercules, which replaced the last Argosy aircraft in 1971, thirty-one Hawker Siddeley Andover and twenty Bristol Britannia and six de Havilland Comet 4 transports; with seventy Westland Wessex and forty Sud-Westland SA.330 Puma helicopters, supplemented by small numbers of Westland-Bell 47G Sioux helicopters at present, and with twenty-five Westland WG.13 helicopters on order for the early 1970s. Training is on de Havilland Canada Chipmunk, BAC 145 Provost, Folland Gnat, Handley Page Hastings, Vickers Varsity and Hawker Siddeley HS.125 Dominie aircraft; with Lightning, Harrier, Hunter, Canberra and Jaguar conversion trainers. Communications aircraft include Beagle Bassets and de Havilland Doves. Two Hawker Siddeley HS.125 jets were introduced in 1971 for VIP transport duties. BAC Bloodhound surface-to-air missiles are deployed.

Fleet Air Arm

Although the history of the Fleet Air Arm in its present form dates only from 1937, in fact the history of British naval aviation is a long one dating from 1909 when the sum of £35,000 for an airship was included in the Naval Estimates. The pre-World War I Royal Navy made gigantic strides towards the present day concept of naval air power with the first take-offs from ships using wooden platforms constructed over the forward gun turrets on HMS *Africa* and *Hibernia* in 1912. That same year the air branches of the Royal Navy and of the Army were merged to form the Royal Flying Corps, but in July 1914 the Royal Naval Air Service was formed, and naval aviation separated from the RFC. Independence was relatively short-lived, however, as in April 1918 the RNAS and the RFC were merged to form the Royal Air Force.

The Fleet Air Arm was actually formed in 1924 as the carrier-borne branch of the Royal Air Force, and it was not until 1937 that full control of naval aviation passed back to the Admiralty once again. In 1924 the FAA operated from five aircraft carriers, HMS *Argus, Eagle, Hermes, Fury* and *Pegasus,* while HMS *Courageous* joined the fleet in 1928, and most battleships and cruisers carried one or two aircraft for spotting and communications purposes. Aircraft in service during the 1920s and 1930s included Avro Bisons, Blackburn Fleet Spotters and Darts, Fairey IIIDs and IIIFs, Flycatchers and Supermarine Seagulls.

In 1937 a massive programme of aircraft-carrier construction was put in hand, but of these ships only HMS *Ark Royal* entered service before the outbreak of World War II, although HMS *Formidable, Illustrious, Implacable, Indefatigable, Indomitable* and

Victorious followed quickly. In spite of operating obsolete aircraft types, the FAA served in every theatre of war. Escort carriers, converted from merchant ships, were evolved for convoy protection. Notable actions included the attack on the Italian Fleet at Taranto in 1940, when Fairey Swordfish from HMS *Illustrious* disabled six enemy battleships. HMS *Victorious* and HMS *Ark Royal* used their aircraft in support of the operation which led to the sinking of the *Bismarck* in 1941. In 1944, Fairey Fireflies wiped out sixty per cent of Japan's aviation spirit production by destroying the refineries at Palembang.

The wartime successes were not without cost, and the FAA lost HMS *Ark Royal* and *Hermes*. At the end of the war the Royal Navy had 1,350 aircraft instead of the 250 in 1939, and fifty-two aircraft carriers instead of the seven in 1939. Wartime aircraft had included the Fairey Swordfish, Barracuda and Fireflies, Hawker Sea Hurricanes, Gloster Gladiators, Grumman Hellcats and Avengers, Chance-Vought Corsairs and Supermarine Seafires.

All of the escort and many of the light fleet aircraft carriers went into reserve during the post-war reduction in strength. Successful deck-landing trials were carried out with the de Havilland Mosquito, Sea Hornet and Vampire, the last being the first jet to land on a ship when landings and take-offs were held aboard HMS *Ocean*. The FAA played a prominent role in supporting ground forces from HMS *Triumph, Theseus, Glory* and *Ocean* during the Korean War, when a Firefly shot down a MiG-15 jet fighter, and in police duties in Malaya. Other post-war developments included the introduction of angled flight decks (which eliminated the need for crash barriers for aircraft overshooting the landing area), mirror landing aids, and the steam catapult – all British inventions which enabled naval aviation to move swiftly, safely and efficiently into the jet age. Re-equipment also took place with the introduction of Douglas AD-4W Skyraider anti-submarine aircraft; Westland Dragonfly helicopters for communications and rescue duties; the Westland Wyvern strike aircraft, which was the world's first operational turboprop aircraft, and was followed by the turboprop Fairey Gannet anti-submarine, airborne-early-warning and carrier onboard-delivery aircraft; and jet de Havilland Sea Venom, Supermarine Attacker and Hawker Sea Hawk fighters.

The world's first rotary-wing aircraft assault on an enemy coast took place in 1956, when British and French forces invaded the Suez Canal zone. Helicopters from HMS *Theseus* were used.

During the late 1950s the FAA received Supermarine Scimitar and de Havilland Sea Vixen fighters. These were followed during the 1960s by Westland Wessex helicopters, which replaced many of the earlier Whirlwind helicopters that had entered service during the mid-1950s, and were later supplemented by the smaller Wasp helicopter for anti-submarine operations from frigates; Mark I and then Mark II versions of the Blackburn Buccaneer carrier-borne bomber; and McDonnell Douglas F-4K Phantom II fighter-bombers. The carrier force was progressively reduced to five during the 1950s, although these were all post-war vessels (HMS *Eagle, Ark Royal, Hermes, Victorious* and *Centaur*), with another two former aircraft carriers, HMS *Albion* and *Bulwark*, as commando carriers operating helicopters. Many frigates, the County class of guided-missile destroyer, and Royal Fleet Auxiliaries, also operated helicopters, as did two assault ships, HMS *Fearless* and *Intrepid*, and plans were set in hand for the three Tiger class cruisers to be converted to helicopter cruisers (HMS *Tiger, Lion* and *Blake*).

The middle and late 1960s saw first the retirement of *Centaur*, then *Victorious* and *Hermes*, while the Tiger class cruisers, of which only two were ever converted to carry helicopters, may go into reserve. HMS *Ark Royal* and *Eagle* were due for withdrawal at the end of 1971, which would have been linked with the end of fixed-wing flying in the Royal Navy, but this has now been changed, with HMS *Eagle* due for retirement at the end of 1971 but with some fixed-wing squadrons remaining while HMS *Ark Royal* remains in service until the mid-1970s. The late 1970s will see a small number of 'through-deck' cruisers enter service which will be able to operate vertical take-off strike aircraft.

Currently the FAA's aircraft include twenty-four McDonnell Douglas F-4K Phantom II fighters in two squadrons; about sixty-four Hawker Siddeley Buccaneer S.2 low-level strike bombers in four squadrons; three squadrons with a total of thirty-six Hawker Siddeley Sea Vixen interceptors; several flights of Fairey Gannet airborne-early-warning aircraft and

Supermarine Scimitar tanker aircraft; sixty Westland Sea King anti-submarine helicopters; 150 Westland Wessex helicopters for anti-submarine and assault duties; ninety Westland Wasp helicopters which during the early 1970s are being replaced by one hundred Westland WG.13 helicopters; plus a few Westland 47G Sioux and some Hiller HT.1 helicopters. In anticipation of the run-down, training has virtually ceased, but de Havilland Sea Devons and Sea Herons, and Percival Sea Prince aircraft operate on communications duties. Two Hawker Siddeley HS.125 jets were obtained in 1971 for V.I.P. transport.

Army Aviation

The British Army's experience of aviation dates, strictly speaking from 1878, but this led to the formation of the Royal Flying Corps and in turn to the Royal Air Force. During World War II, twelve AOP squadrons were operated with Army and RAF personnel, and during the 1950s, some of these continued with the addition of liaison flights. A large number of pilots had been trained during the war for troop-carrying Horsa and Hamilcar invasion gliders, and many of these were retrained for service with the liaison flights. The formation of the Army Air Corps in 1957 led to the gradual disappearance of RAF personnel.

At present, although the AAC exists, it is not on a permanent basis and most flying and ground personnel are members of other corps or regiments. Duties of the AAC and the other elements within the overall organization known as Army Aviation (which has its own headquarters) consist mainly of AOP and reconnaissance, fire support including anti-tank operations using SS.11 wire-controlled missiles fired from Westland Scout helicopters, liaison, communications and light transport. Apart from the headquarters, organization, as formulated in 1969, consists of a division between the British Army of the Rhine, Strategic Command in the United Kingdom, and units at Hong Kong, the Persian Gulf and the Far East, although the future of the last two is uncertain depending on the exact strength and nature of the forces to be deployed in these areas in future. Each of the 'theatres' is further broken down into divisions, each with an aviation regiment, with an HQ squadron and brigade squadrons.

Current equipment includes 150 Westland Scout helicopters which are due to be replaced during the early 1970s by 250 Westland WG.13s; 175 Westland 47G Sioux helicopters, due to be replaced during the early 1970s by three hundred Westland-Sud SA.340 Gazelles; fifteen Sud Alouette IIs; and a number of Bell G4 training helicopters, which are operated by a contractor, Bristow Helicopters, as are a number of de Havilland Canada Chipmunk trainers. Forty de Havilland Canada DHC-2 Beaver utility transports are also operated.

UNITED STATES OF AMERICA

United States Air Force

Although balloons were used for observation purposes in the 1860s, during the American Civil War, the history of American military aviation can be best dated from 1892, when a balloon section was formed within the United States Army for use during the Spanish-American War. The balloon section saw service in Cuba in 1898. The US Army supported several unsuccessful attempts to build workable aircraft shortly after the turn of the century, but refused offers of the Wright Brothers aircraft. In 1907 the Aeronautical Division of the Signal Corps took over responsibility for military aviation, and the following year this received a Wright Brothers aircraft and a dirigible. By 1911 four Wright and two Curtiss aircraft were being operated on training and reconnaissance duties, but this number was down to five aircraft by 1914 when World War I broke out in Europe. However, that same year the Signal Corps Aeronautical Section was formed, becoming the US Army Aviation Section in 1916. The United States entered World War I in 1917 on the side of the Allies, and by this time some two hundred aircraft were in service, with Curtiss, Martin, Standard and Wright machines predominating.

Ambitious wartime expansion schemes were put into effect, mainly dependent on European designs built under licence, including Bristol F.2B 'Brisfit' and Spad fighters, with some D.H.4 bombers, while a few aircraft were bought to supplement these types. All in all, about 5,000 French-built aircraft, including those of Breguet, Caudron, Farman, Morane-Saulnier, Nieuport, licence-built Sopwith, Spad and Voisin manufacture, and nearly three hundred British-built aircraft of Avro, Airco, Bristol and Sopwith manufacture were operated by the US Army during the war. Rapid expansion of the US Army Aeronautical Section was made possible by the return of many Americans who had been flying with the other Allied air arms pending the United States' entry into the war.

The post-war strength was settled at 2,500 aircraft in 1920, but few new aircraft were forthcoming. Attempts to form a separate air service on the RAF pattern failed, and in 1920 the US Army Air Service was formed with twenty-seven combat squadrons of eighteen aircraft each. The USAAS had grown to forty-eight squadrons by 1923, with the D.H.4B bomber as its mainstay. To maintain morale and gain publicity, long-distance flights and attacks on abandoned warships were made. This was followed in 1926 by yet another change of name, to the US Army Air Corps. Aircraft then in service included Curtiss PW-8, P-1 and P-2 fighters and O-1 AOP aircraft, Martin MB-2 and licence-built D.H.4M bombers, Douglas O-2 AOP aircraft, Consolidated PT-1 trainers and Douglas C-1 transport aircraft. The depression delayed a programme of expansion which was due to start in 1926.

However, new aircraft did arrive during the late 1920s and early 1930s, including 150 Curtiss A-3 Falcon ground-attack aircraft; twelve Curtiss B-2, and numbers of Keystone LB-5, LB-6, LB-7, B-3, B-4 and B-6 bombers; twelve Fokker F.VII and thirteen Ford C-3 and C-4 Trimotor transports. These were followed by Berliner-Joyce P-16, Boeing P-12B and Curtiss P-16, fighters. There were also four balloon squadrons still in service during this period.

Although somewhat belated, the US Army Air Corps was gradually reaching the size and equipment quality of the air arm of a major power. The 1930s saw this process accelerate further, with still more new aircraft, including the Boeing P-26A fighter of 1933 and the Martin B-10 bomber of the following year; while in 1935 a series of accidents following the transfer of mail services from the airlines to the USAAC led to an investigation resulting in a degree of semi-autonomy for the service.

The first of fifty Consolidated P-30A and seventy-six Seversky P-35 fighters, thirty-two Martin YB-12 and 250 Northrop A-17 bombers and attack aircraft respectively, and ninety Douglas O-46A AOP aircraft in 1935, were followed in 1936 by the Douglas B-18, the bomber version of the famous C-47 Dakota transport. Meanwhile the Army and the Navy contended over the responsibility for maritime-reconnaissance duties.

After the Munich crisis of 1938, expansion was speeded up considerably. Very large orders were given, particularly after World War II started in Europe in 1939. Amongst the aircraft ordered were additional Boeing P-26, Seversky P-35 and Curtiss P-36 fighters; Northrop A-17 and Curtiss A-12 and A-18 ground attack aircraft; Martin B-10 and B-12, Boeing B-17 Fortress and additional Douglas B-18 bombers; Douglas OA-3 and OA-4, Sikorsky OA-8 and Grumman OA-9 amphibians; Bellanca C-27, Douglas C-33 and C-39, Lockheed C-36 and C-40 transports; North American O-47 and Douglas O-38, O-43, and O-46 AOP aircraft; and North American AT-6 and BT-9, Consolidated PT-11, Stearman PT-13 and Seversky BT-8 trainers.

The next generation of aircraft benefited considerably from service with the British and French armed forces before full-scale entry into US Army and US Navy service. These included Lockheed P-38 Lightning fighters, and Consolidated B-24 Liberator, Martin B-26 Marauder and North American B-25 Mitchell bombers. In 1941 the USAAC became the United States Army Air Force, with its four air forces, 1st and 2nd in the North, 3rd and 4th in the South, deployed for the defence of the United States. By the time the Japanese attacked Pearl Harbor on 7 December 1941, bringing the United States into World War II, the USAAF had some 2,500 aircraft, of which the latest were the Bell P-40, Curtiss P-40, Republic P-43 and P-47 (later F-47 when the classification was changed from P–pursuit to F–fighter) Thunderbolt, and North American P-51 Mustang fighters; Douglas A-20 Boston and A-24, and Curtiss A-25 attack aircraft; Beech C-45, Curtiss C-46 Commando and Douglas C-47 Dakota, C-53, C-54 and Lockheed C-59 and C-60 transports. Most of the USAAF aircraft in Hawaii and the Philippines were destroyed on the ground by air attack, and the Japanese subsequently occupied the Philippines.

From 1942 onwards, the USAAF operated in increasing strength in North Africa, and from the British Isles over Europe, from Hawaii over the Pacific, and from India over China and Burma. Few combat aircraft were left in the United States. In 1943 the USAAF took part in the invasion of Sicily, and the following year took part in the Normandy landings. In each case a large part of the effort consisted of transport aircraft operation and towing troop-carrying gliders. In Europe the USAAF policy of daytime bombing raids led to heavy losses of the Boeing B-17 Fortress and Consolidated B-24 Liberator bombers when faced by the Luftwaffe's Messerschmitt Bf 109F fighters. The difficulty lay in providing effective fighter cover over such long ranges, but further deliveries of the Lockheed P-38 Lightning and North American P-51 Mustang helped to remedy this state of affairs. For the most part, aircraft in service towards the end of the war were developments of those introduced during the early 1940s, but there were one or two significant new types, including the Northrop P-61 Black Widow night-fighter, and the Boeing B-29 Superfortress bomber. (It was this last aircraft which was to drop the first atomic bombs on Japan, on Hiroshima on 6 August 1945, and Nagasaki on 9 August 1945, bringing the surrender of Japan and an end to World War II.)

The USAAF ended the war as the world's largest air force, with something like 60,000 aircraft. In 1946 it was reorganized into Air Defence, Tactical and Strategic Air Commands, with Air Material, Air Proving, Air Training and Air Transport Commands providing support. The long-cherished ambition to become a separate service was realized in 1947 when the USAAF became the United States Air Force. That same year saw the USAF receive its first jet fighter, the Lockheed P-80 Shooting Star, which almost immediately had its designation changed to F-80 when the F–fighter was adopted instead of the P–pursuit definition. The Republic F-84 Thunderjet supplemented the Shooting Star in 1948. Meanwhile the Boeing B-50, a higher-powered version of the Superfortress, and the Convair B-36, a six-engined bomber designed to bomb Germany if Britain had fallen, had both been introduced in 1947. Many B-29 Superfortresses were converted to the tanker role for inflight refuelling, a British invention.

However, the return of peace was relatively shortlived. In 1948, Fairchild C-82, Douglas C-47 Dakota and C-54 Skymaster transports played their part in the Allied airlift of supplies to West Berlin, which had been blockaded by Russian forces. In 1950 a war broke out in Korea but, prior to this, the North American F-86 Sabre and the Lockheed F-94 Starfire fighters had entered service followed by North American B-45 Tornado light jet bombers, and these aircraft saw service with the United Nations Forces in Korea. The Air National Guard was called up for the Korean War, doubling the USAF's strength. The Beech T-34 Mentor, North American T-28 Trojan and Lockheed T-33A trainers also entered service at this time, the last-mentioned being a jet trainer.

The Korean War lasted from 1950 to 1953, during which time USAF units operating under United Nations auspices participated in some of the first jet fighter battles when Lockheed P-80C Shooting Stars faced Mikoyan MiG-15s. The end of the Korean War was brought about by American air supremacy in the area.

The first heavy jet bomber, the Boeing B-47 Stratojet entered service with Strategic Air Command in 1951, while in 1955 the eight-engined Boeing B-52 Stratofortress replaced the Convair B-36. Both types remain in service, as does one of the most useful night-fighter, reconnaissance and bomber types built during this period, the Martin B-57, a licence-built version of the English Electric Canberra jet bomber. Other aircraft of this period included the North American F-100 Super Sabre, the McDonnell F-101 Voodoo, Northrop F-89 Scorpion fighters and interceptors; Douglas B-66 Destroyer bombers; Lockheed RC-121 Constellation airborne-early-warning aircraft; Douglas C-124 Globemaster and C-133 Liftmaster and Fairchild C-119G Packet and C-123 Provider transports; followed during the late 1950s by Convair F-102 Delta Daggers and F-106 Delta Darts, and Lockheed F-104 Starfighter interceptors; the Convair B-58A Hustler supersonic bomber and the Lockheed C-130 Hercules transport. The late 1960s saw the Northrop F-5 Freedom Fighter tactical fighter-bomber, the McDonnell F-4 Phantom II fighter-bomber, LTV-A-7D attack aircraft, the General Dynamics F-111 variable-geometry strike aircraft, the Lockheed C-141 Starlifter and the C-5 Galaxy transports all enter service. Vertol and Sikorsky helicopters were introduced during the 1950s.

Military aid has also been provided to a number of countries within the Organization of American States, the North Atlantic Treaty Organization, and the South-East Asia Treaty Organization; while the United States is the major partner in a defence pact with Australia and New Zealand. During the 1960s a major commitment of the USAF was in Vietnam, supporting ground forces fighting Communist infiltrators from the north, and also undertaking bombing raids against targets in North Vietnam.

Currently the USAF has 810,000 men, and operates eleven squadrons of Convair F-106A Delta Darts, one F-102A Delta Dagger squadron, and three McDonnell Douglas F-101B Voodoo squadrons in the interceptor role. In addition there are some 2,000 McDonnell Douglas F-4C/D/E Phantom II fighter-bombers, including a couple of hundred RF-4C Phantom II reconnaissance-fighters, while Republic F-105 Thunderchiefs and North American F-100D Super Sabres are being replaced by large numbers of Northrop F-5A Freedom Fighters, two hundred or so General Dynamics F-111A fighter-bombers and about six hundred Ling-Temco-Vought A-7D Corsair II attack aircraft; there are also some thirty squadrons with a total of some four hundred Boeing B-52G/H Stratofortress, and perhaps sixty-four General Dynamics FB-111 bombers. There are a fairly large number of Cessna A-37 armed-trainers and O-2 forward control aircraft and gunship versions of the Douglas C-47 (AC-47), Fairchild C-119 (AC-119), and C-123 (AC-123), and Lockheed C-130 (AC-130); three Lockheed EC-121 airborne-early-warning aircraft squadrons; and several hundred Boeing KC-135 tanker aircraft. Transport aircraft include three hundred Lockheed C-130 Hercules, approximately three hundred C-141A Starlifter, eleven VC-140 JetStars, and about one hundred C-5A Galaxy, a decreasing number of Douglas C-47 Dakota and C-117 Skytrain, C-133 Cargomaster, and nine C-9A, and thirty Boeing C-135 aircraft, with numerous light transport and communications types. A large number of Lockheed C-130 variants are in use in weather and rescue units, which also use Kaman HH-43B Huskie, Sikorsky S-61 and HH-53 helicopters, and Grumman HU-16 Albatross amphibians. Training types include Beech T-34A Mentor, North American T-28 Trojan, Cessna T-41 (normally operated by contractors), and T-37, and Northrop T-38A aircraft.

About fifteen Lockheed S-71C and thirty Lockheed U-2 aircraft are used on reconnaissance duties, as are some Martin B-57 conversions, while McDonnell Douglas EB-66 Destroyers operate electronic countermeasures and reconnaissance flights. Most surface-to-air missiles are US Army operated, but some Bomarc missiles are deployed in the USAF. An anti-missile missile system, Safeguard, is under development at present, while the McDonnell Douglas F-15 fighter should enter service during the mid-1970s. Air National Guard (reserve) units are re-equipping with Cessna A-37 armed-trainers; North American F-100 Super Sabre and Republic F-105 Thunderchief fighter-bombers; and Lockheed C-130A Hercules transports, replacing older aircraft of early 1950s vintage.

United States Navy and Marine Corps

Funds were first voted for United States Navy aircraft in 1911, and a Wright and two Curtiss aircraft were soon in service. Two years later, a total of eight aircraft were being operated, including a few seaplanes which operated from the battleships USS *Birmingham* and *Mississippi* against Mexico. At the start of America's participation in World War I in 1917, there were twenty-one aircraft in USN service, but by the time the war ended there were more than 1,000 seaplanes and flying-boats and 250 landplanes in USN and United States Marine Corps service. During the early part of the war, drastic shortages of aircraft had resulted in personnel serving with other Allied air arms, but later some anti-submarine and bombing missions were mounted. Curtiss R-6 and HS2-2 seaplanes and H-16 flying-boats predominated, but there were also fairly large numbers of Naval Aircraft Factory and foreign products.

Post-war organization and equipment of the USN and USMC air arms was based on D.H.4B, Curtiss JN-4H, Martin MBY and Morse M.1 aircraft, Curtiss N-9 and R-6, and Boeing CL-4 seaplanes, with Curtiss N.C.1, N.C.3 and N.C.4, F.5L and F.5, Aeromarine 40L and Standard HS.2L flying-boats. A collier was converted to become America's first aircraft carrier, the USS *Langley* in 1922, and Sopwith Scout fighters were operated from platforms fitted over the forward gun turrets of several battleships.

During the early 1920s, Naval Aircraft Factory PT and Davis-Douglas aircraft were obtained, along with sixty Vought UO-1, sixty Martin MO-1 and some Douglas DT aircraft – all of these types could be operated from the USS *Langley* while a variety of British and European aircraft were obtained for evaluation. In 1925, Curtiss F-4C fighters, seventy-five Martin SC-1 and Douglas T2D torpedo-bombers and Vought O2U reconnaissance aircraft entered service. Throughout the period the USN maintained a strong interest in balloons and dirigibles.

Two more aircraft carriers, which had been converted from battle-cruisers while under construction, the USS *Saratoga* and *Lexington*, entered service. By this time there were also some twenty battleships each with three catapult-operated aircraft, and ten cruisers with two seaplanes each. During the early 1930s a fourth aircraft carrier, the USS *Ranger*, entered service, while aircraft at that time included Boeing F3B-1 and F4B-2 fighters, Chance-Vought O2U-4 and O3U-2 Corsair and Curtiss O2C-1/2 observation aircraft, and Consolidated P2Y-1 flying-boats, followed by Curtiss F8C-4 Helldiver dive-bombers, and a number of Curtiss F9C-2 Sparrowhawk fighters, which operated from airships as an experiment. More than two hundred aircraft entered service at this time, which also saw the first USMC carrier-based operations.

Two more aircraft carriers entered service in 1937, the USS *Yorktown* and *Enterprise*, and were followed by the smaller USS *Wasp*. USN and USMC aircraft at this time included the Boeing F4B-2 already mentioned; Curtiss BF2C-1 Hawk and Grumman FF-1, F2F-1 and SF-1 fighters; Chance-Vought SB2U-1 reconnaissance-bombers; Martin BM-1 and BM-2 dive-bombers; Curtiss SOC-1 and O3C-1 and Chance-Vought O3U-6 Corsair observation aircraft; Consolidated P2Y-1 and P3Y-1 flying-boats; Grumman JF-1 amphibians; and Stearman NS-1 trainers. By 1940 the USN and USMC had a total strength of 3,000 aircraft. The new aircraft of the early 1940s included many due to become famous during World War II, which had already started in Europe: Brewster F2A, Vought F4U Corsair, and Grumman F4F Wildcat and F6F Hellcat fighters; Douglas SBD Dauntless and Curtiss-Wright SB2C Helldiver dive-bombers; Douglas TBD Devastator and Grumman TBF Avenger torpedo-bombers;

Vought Sikorsky OS2U Kingfisher observation aircraft; Consolidated PBY-1 Catalina amphibians and Martin PBM Mariner flying-boats. Starting in 1941, a programme of converting merchant ships to aircraft carriers for escort duties was to result in more than one hundred such conversions for the United States Navy and the Royal Navy, while another full-sized carrier, the USS *Hornet*, also entered service that year.

Eleven full-sized aircraft carriers were under construction at the time of the Japanese attack on Pearl Harbor, on 7 December 1941, which brought the United States into World War II. Shortly before the attack, the deteriorating international situation had led to the United States Coastguard Service being transferred to the Navy – a normal wartime practice – bringing with it General PJ-1, Hall PH-2 and PH-3, and Viking OO-1 flying-boats; Consolidated PBY-5 Catalina, Grumman JF-2, JRF-2 and J4F-1 Goose and Douglas RD-4 amphibians; Curtiss SOC-4, Fairchild JK-1 and JK-2, Lockheed R-30 and R-50 and Waco JRW-1 aircraft. However, when the Japanese attacked Pearl Harbor sinking eight battleships, the Pacific Fleet's aircraft carriers were not in port.

The USN and USMC fought mainly in the Pacific during World War II, notable actions being the Battle of the Coral Sea, the Battle of Midway, from which the Japanese fleet was unable to recover, taking part in the invasion of Gaudalcanal, and carrier-based bombing raids on Japan itself. But there were also escort duties with the North Atlantic convoys, while the USS *Wasp* ferried aircraft to Europe, and also from the United Kingdom to Malta. The losses of the USS *Langley, Wasp, Lexington, Hornet* and *Yorktown*, plus several escort carriers, was countered by a massive carrier construction and conversion programme, so that by the end of the war no less than ninety-eight carriers of all types, including six USMC carriers, were operational. New aircraft types in service during the war were few, and mainly land-based types, including the Boeing B-17 Fortress and B-29 Superfortress, North American B-25 Mitchell, and Lockheed PV-1 and PV-2 Harpoon bombers, mainly for maritime-reconnaissance patrols. There were also numbers of transport aircraft, including Curtiss C-46 Commando, Douglas C-47 Dakota and Lockheed Lodestars.

A rapid reduction in wartime strength after the end of the war meant that the USN aircraft were cut from some 40,000 to 10,000.

However, the reduced numbers were of modern aircraft, as types which had been developed too late to take part in the war started to enter service in quite large numbers, including Grumman F7F Tigercat, F8F Bearcat, and Ryan FR-1 Fireball fighters; and Douglas A-1 Skyraider and Martin AM-1 Marauder ground-attack aircraft; while Lockheed P2V Neptune landbased maritime-reconnaissance aircraft eventually took over from the Harpoons, Fortresses and Superfortresses. The USN's first jet arrived, the McDonnell FH-1 Phantom. The transport fleet continued to develop with deliveries of Fairchild C-119 Packets, Douglas C-54 Skymasters and transport versions of the Grumman VF-1 Albatross amphibian and Martin JRM Mars flying-boat, which were also used in air-sea rescue versions. The lack of sufficient carrier-borne jets meant that, for a while, land-based Lockheed F-80 Shooting Stars had to be operated. Between 1948 and 1956 a Military Air Transport Command was formed which took over gradually all USAF, USN, USMC and US Army transport aircraft.

Eventually the long-awaited carrier-borne jets arrived, including Grumman F9F Panther, McDonnell F2H Banshee, Chance-Vought F6U Pirate and North American FJ-1 Fury (USN and USMC versions of the F-86 Sabre) fighters, while Sikorsky S-51 helicopters were obtained for plane-guard and communications duties. At the start of the Korean War, the USN had some thirteen aircraft carriers, and further cuts in USN and USMC strength were postponed as a result of the war. Many reserve personnel had to be called up between 1950 and 1953 while the Korean War lasted. The main activities of the USN and USMC aircraft were in providing close support for ground forces.

During the Korean War, the North American AJ-1 Savage attack-bomber entered service, with a number of variants and developments of existing aircraft, including airborne-earlywarning Skyraiders. The early 1950s also saw Douglas F4D-1 Skyray, Grumman F9F-6 Cougar and McDonnell F3H Demon fighters enter service, with land-based Lockheed R70 (maritime-reconnaissance Super Constellations) and Fairchild R49 aircraft. Lockheed TV-2 jet trainers and North American T-28B Trojan trainers were also obtained.

New aircraft carriers included the USS *Forrestal* and *Saratoga,* equipped with all the latest aids developed and pioneered by the Royal Navy, although ironically it was the American ships that were first provided with all of these. Aircraft of the late 1950s included Chance-Vought F8U-1 Crusader and F7U-3M Cutlass fighters and the Douglas A3D-2 Skywarrior carrier-borne bomber. These in turn were followed by the very successful McDonnell Douglas F-4 Phantom II and A-4 Skyhawk fighter-bombers, the Ling-Temco-Vought A-7E Corsair II attack aircraft and land-based Lockheed P-3 Orion maritime-reconnaissance aircraft of the 1960s. The Grumman S2F-1 Tracker anti-submarine aircraft, WF-2 Tracer airborne-early-warning aircraft and Trader carrier onboard-delivery aircraft of the late 1950s started to be replaced on the larger carriers by new aircraft, including the Grumman E-2 Hawkeye, while Skywarriors are now used as tankers. Grumman A-6A Intruders replaced some of the Skyhawks during the latter part of the 1960s, while Grumman F11F Tigers and Super Tigers of the late 1950s and early 1960s have now disappeared. The USMC are receiving 112 Hawker Siddeley Harrier AV-8A vertical take-off fighters, with deliveries started at the end of 1970, and it is likely that in the not-too-far-distant future that the USN will receive higher-powered developments of this aircraft.

Currently the USN operates fifteen attack carriers, some of which are nuclear-powered, with up to one hundred aircraft each, and seven anti-submarine carriers with about fifty aircraft each, and the total number of USN aircraft is about 8,000, while the USMC has about 1,200 aircraft. Interceptors include McDonnell Douglas F-4 Phantom IIs and Ling-Temco-Vought F-8 Crusaders; while Phantom II, McDonnell Douglas A-4 Skyhawks, Grumman A-6A Intruders and LTV A-7E Corsair IIs are used for attack duties; North American RA-5C Vigilante and LTV RF-8G Crusaders are used on reconnaissance duties; with Douglas KA-/B Skywarrior tanker aircraft. Grumman E-2A Hawkeye airborne-early-warning aircraft are carried on the attack carriers, while Grumman S-2E Trackers are used for maritime-reconnaissance from the anti-submarine carriers. A number of Grumman Tracers remain in AEW service on the smaller carriers, while Grumman Traders are used still for carrier onboard delivery (on tactical transport) duties, in addition to which there are twenty-four squadrons with a total of some two hundred shore-based Lockheed P-3 Orion maritime-reconnaissance aircraft. Five squadrons of transport and communications Douglas C-118 Liftmaster, Lockheed C-130 Hercules, and Grumman C-1A and C-2A Greyhound COD aircraft are still left with the USN. Helicopters include Sikorsky SH-3A/D Sea King antisubmarine and SH-34G, Kaman UH-2 Seasprite search and rescue types, and Boeing-Vertol UH-46A transport helicopters; with training on Beech T-34 Mentor and TC-45, North American T-28B Trojan and T-2A/B Buckeye, Douglas TA-4F, Lockheed T-1A Seastars, and Bell TH-57A helicopters. A typical attack carrier would operate almost any combination of the above, but anti-submarine carriers by their size are restricted to a standard group of two squadrons with a total of twenty Grumman S-2 Trackers, sixteen Sikorsky SH-3D Sea Kings in one squadron, four Grumman E-1B Tracers, two Kaman UH-2 Seasprites for plane-guard duties, and four Douglas Skyhawks for fighter protection.

USMC aircraft include fourteen McDonnell Douglas F-4 Phantom II fighter squadrons, twelve Grumman A-6 Intruder squadrons, and three RF-4B Phantom II reconnaissance squadrons. There are 112 Hawker Siddeley Harrier AV-8A vertical take-off fighters entering service; and there are three squadrons of Lockheed KC-130F Hercules tanker and assault aircraft; fourteen Sikorsky UH-34D and Boeing-Vertol CH-46A transport helicopter squadrons; 106 Sikorsky CH-53A Sea Stallion, twenty-four Bell UH-1E and thirty-eight AH-1G Iroquois helicopters; with 124 North American OV-10A Broncos for armed-reconnaissance and target-towing duties.

The United States Coastguard operates a large number of helicopters and Grumman HU-16 Albatross amphibians on search and rescue duties, in which they are supported by Lockheed C-130 Hercules aircraft.

Grumman F-14A Tomcat variable-geometry aircraft are being developed for the USN's use in the late 1970s. A three-engined

Sikorsky CH-53 helicopter development is in hand. Two 94,000-ton nuclear-powered aircraft carriers are under construction – USS *Nimitz* and *Dwight D Eisenhower*.

United States Army Aviation

Although the USAAF was under overall Army command until it became the USAF in 1947, after 1938 USAAF aircraft were no longer under the control of local Army commanders. This state of affairs led the Army in 1942 to purchase Piper L-4 Cub, Stinson L-5 Sentinel, Vultee-Stinson L-1, Taylorcraft L-2 and Aeronca L-3 AOP aircraft, about 3,000 in all and probably the largest ever initial equipment purchase for an air arm. Additional deliveries of these aircraft and later models built the strength of US Army Aviation up to a maximum of some 4,500 aircraft during World War II.

During the war the first helicopters, Sikorsky R-4, R-5 and R-6, entered service; and afterwards Sikorsky H-5 and H-19 Chickasaw, and Bell H-13 Sioux helicopters were used extensively. New fixed-wing aircraft were also introduced, of which the most numerous was the Cessna L-19 Bird Dog, a type which first appeared in 1950. By the middle 1950s, about 1,800 each of fixed-wing and rotary-wing aircraft were in US Army service, operating on AOP, liaison, communications, transport and reconnaissance duties. The latter part of the 1950s saw Sikorsky H-34A Choctaw and H-37A, Vertol H-21C and H-25C, Hiller H-23C and additional Bell H-13 Sioux enter service. However, fixed-wing aircraft in the form of 370 de Havilland Canada L-20 (DHC-2) Beaver were acquired and, later, numbers of U-1A (DHC-3) Otter light transports; with Beech L-23 Twin Bonanza, Cessna L-27A (310) and Aero L-26 Commanders for communications duties. Deliveries since then have included the Bell UH-1A Iroquois and the Boeing Vertol CH-47 Chinook helicopter. Armed escort and gunship helicopters, such as the Bell AH-1G Hueycobra, are also in use; and during the 1960s large numbers of aircraft have been operated with ground forces in South Vietnam and (in 1970) Cambodia against Communist North Vietnamese infiltrators.

Currently, with some 10,000 helicopters, the United States

Army is the world's largest helicopter operator, and there are in addition another two thousand five hundred to three thousand fixed-wing aircraft. Helicopters include two thousand Bell OH-13 Sioux, 2,200 OH-58A JetRanger, four thousand UH-1A/B/D Iroquois, 750 AH-1G Hueycobra, 1,300 Hughes OH-6A, six hundred Boeing-Vertol CH-47A, ninety Sikorsky CH-37B and fifty-six CH-54A. Fixed-wing aircraft include Beech C-45s, Douglas C-47 Dakotas, Lockheed VC-140 JetStars, with eight hundred de Havilland Canada Beaver and two hundred Otter, one thousand Cessna Bird Dog and three hundred Grumman OV-1 Mohawk types. Training is on 250 Cessna T-41B and sixty Beech T-42As, with eight hundred Hughes TH-55A and four hundred Bell TH-13T helicopter trainers. Future equipment will probably include Sikorsky S-67 Blackhawk utility and attack helicopters.

URUGUAY

Uruguayan Air Force *Fuerza Aérea Uruguaya*

Uruguayan military aviation dates from 1916, when three Morane-Saulnier aircraft were acquired for a newly-established Department of Military Aviation. In 1919 four Avro 504Ks were bought, and in 1921 six Nieuport 33 trainers. The air arm was named the Aeronáutica Militar, and during the 1920s and 1930s aircraft operated included Ansaldo A.300 fighters, Potez XXVA and Romeo Ro 37 reconnaissance-bombers, a Farman F.190 and a Stinson AOP aircraft, a Breguet Br.19 transport, Waco D-7 general-purpose aircraft, and de Havilland Tiger Moth trainers.

Although Uruguay did not play an active part in World War II, bases were put at the disposal of United States forces, and in exchange a considerable amount of military aid was received, including Grumman F6F-5 Hellcat fighter-bombers; TBM-1C Avenger torpedo-bombers and J4F-1 Gosling amphibians; North American B-25J Mitchell bombers; AT-6 and SNJ-4 trainers; Chance-Vought OSU-3 Kingfisher AOP aircraft; Curtiss C-46 Commando and Douglas C-47 Dakota transports; Beech T-11B, Fairchild PT-23A and PT-26 trainers. The post-war period saw

North American F-51D Mustang fighter-bombers enter service, to be replaced during the early 1960s by Lockheed F-80C Shooting Star jet fighter-bombers and T-33A jet trainers. The present title was adopted in 1952; and in 1948 Uruguay had become a member of the Organization of American States.

Currently the FAU has some 1,600 men and operates ten Lockheed F-80C Shooting Star fighter-bombers; five Curtiss C-46 Commando, thirteen Douglas C-47 Dakota, two Beech Queen Air, two Fokker F-27M Troopship, and two Fairchild-Hiller FH-227B, and a de Havilland Canada DHC-2 Beaver as transports; one Bell OH-13G Sioux and two Hiller H-23F helicopters; twenty North American T-6G Texan, ten Beech T-11B and six Lockheed T-33A trainers.

Naval Aviation *Aviación Naval*

This force dates from 1920, and currently operates five Grumman F6F-5 Hellcat fighters and three S-2A Tracker anti-submarine aircraft; two Martin PBM-5 Mariner flying-boats; four Hiller UH-12 and two Bell 47G Sioux helicopters; and a number of Beech T-34 Mentor and North American T-6 Harvard trainers; with Piper Cubs for liaison and communications duties.

VENEZUELA

Venezuelan Air Force *Fuerzas Aéreas Venezulanas*

Venezuelan military aviation dates from 1920 with the formation of a Venezuelan Military Air Service, which received its first aircraft the following year: Caudron G.III and G.IV and Farman F.40 types, of which there were more than sixteen altogether. The Venezuelan Navy also formed an air arm during the early 1920s. Little further progress was made by either air arm until 1936, when the Military Air Service became the Military Aviation Regiment and a number of Consolidated aircraft were bought. The next few years marked something of an expansion programme, with the MAR operating a domestic airline service, and Fiat C.R.32 fighters and B.R.20 bombers entering service after an Italian Aviation Mission had visited Venezuela.

During World War II, Venezuela stayed on the side of the Allies, ready to prevent any expansion of Axis influence to the Caribbean, and permitted American bases to be in Venezuela. In return, military aid was received during this period, including Republic F-47D Thunderbolt fighters, North American B-25J Mitchell bombers and NA-16-3 and T-6, Beech T-7 and T-11B trainers. After the war ended, Venezuela became a member of the Organization of American States, and the two air arms were amalgamated to form a separate service, the Fuerzas Aéreas Venezulanas.

The new air force was modernized during the early 1950s with de Havilland FB.5 Vampire jet fighter-bombers and English Electric Canberra B.2 jet bombers, followed in 1955 by de Havilland Venom FB.4 fighter-bombers and Vampire T.55 trainers, North American F-86F Sabre fighter-bombers, additional Canberras and Fairchild C-123B Provider transports. In 1967 ex-Luftwaffe F-86K Sabres were obtained.

Currently the FAV has 9,000 men and operates eighteen North American F-86F and forty F-86K Sabre, with a small number of de Havilland Vampire FB.5 and Venom FB.4 fighter-bombers; fifteen BAC Canberra B.2 and PR.3 bombers; eighteen Fairchild C-123B Provider, a Hawker Siddeley 748, and twenty or so Douglas C-47 Dakota and C-54 Skymaster transports; four Sikorsky UH-19 and two S-51, twenty Sud Alouette III and six Bell 47G/J helicopters; fifteen BAC Jet Provost T.52, thirty Beech T-34 Mentor, a number of North American T-6G Texan, and Beech T-7 and T-11 trainers. Twelve Cessna 182 Skylane communications aircraft were ordered in 1971, along with four Lockheed C-130H Hercules transports.

VIETNAM (SOUTH)

Vietnamese Air Force

Vietnam became an independent republic after the French withdrawal from Indo-China in 1954, and was divided into the

North and South states by a Geneva Conference, with unification originally planned for 1956 but postponed indefinitely since the necessary elections have still to take place in the North. South Vietnam took over the Vietnamese Air Force from the French, who had been responsible for its formation in 1951. Initially, Dassault M.D.315 Flamant communications aircraft and Morane-Saulnier M.S.500 trainers were operated, and after the French withdrawal Grumman F8F-1 Bearcat fighter-bombers, Douglas C-47 Dakota transports and North American T-6G Texan trainers were added to the strength.

In 1957, Communist forces crossed from the North to the South and started the intense guerrilla warfare which continues to the present time, with American, Australian, New Zealand, South Korean and Nationalist Chinese contingents also fighting in South Vietnam. American military aid has also been given to the South Vietnamese, including additional C-47s and T-6Gs, Northrop F-5A, Cessna A-37 and Douglas A-1 Skyraider fighter-bomber and attack aircraft, numerous helicopters (with Bell UH-1H Iroquois and Sikorsky CH-34 types predominating), de Havilland Canada DHC-2 Beaver transports, Cessna O-1 Bird Dog AOP aircraft, and training types.

Currently the South Vietnamese Air Force is 23,000 men strong and operates one squadron with seventeen Northrop F-5A Freedom Fighters, three squadrons with a total of sixty Cessna A-37B; and three squadrons with another sixty Douglas A-1 Skyraiders, all operating in the tactical fighter-bomber role; eighty Cessna O-1 Bird Dog AOP aircraft, with light armament; forty-five Douglas C-47 Dakota, fifty-five Fairchild C-119G Packet, de Havilland Canada DHC-4 Caribou and seven DHC-2 Beaver transports; 130 Bell UH-1H Iroquois, sixteen Boeing CH-47 Chinook, a small number of Sikorsky SH-34 Choctaw, and two each of Sud Alouette II and III helicopters; with twenty-five Cessna U-17As for training and liaison.

PEOPLE'S REPUBLIC OF VIETNAM (NORTH)

Vietnamese People's Air Force

After the French withdrawal from Indo-China in 1954, the Communists took control of the northern part of Vietnam. In 1957, North Vietnamese guerrillas started the series of incidents in the South which have now grown into what is known as the Vietnamese War. During this period the Vietnamese People's Air Force came into being, with Russia and China competing to give military aid.

Currently the strength of the People's Air Force is 4,500 men, with thirty Mikoyan MiG-21PF interceptors; sixty MiG-17 and forty MiG-15 fighter-bombers; ten Ilyushin Il-28 bombers; eight Antonov An-2, and three An-24, forty Ilyushin Il-14 and six Il-12, with three Lisunov Li-2 (C-47) transports; some Mil Mi-1, about thirty Mi-4 and twelve Mi-6 helicopters. SA-2 Guideline surface-to-air missiles are deployed.

YEMEN

Yemen Republican Air Force

The Yemen Republic Air Force was formed during the mid-1950s as the Yemen Air Force, with a small number of aircraft of Cessna, Aero Commander and North American manufacture, and a CCF Norseman. During the late 1950s and early 1960s, aid started to arrive from Russia, Egypt and Czechoslovakia, including thirty Ilyushin Il-10 ground-attack aircraft, Mil helicopters and Yakovlev Yak-11 trainers. Training of Yemeni personnel was not successful, and Soviet and Egyptian personnel have had to do most of the flying and maintenance duties. From the mid-1960s onwards, the Yemen was occupied by up to 50,000 Egyptian troops who supported the overthrow of the Imam by Republican forces. Sporadic fighting between the Republican Government and tribesmen supporting the Imam has continued since.

Current equipment includes twelve Mikoyan MiG-17 fighter-bombers and a similar number of Ilyushin Il-28 bombers, with a few Douglas C-47 Dakota and Ilyushin Il-14 transports, Mil Mi-4 helicopters and Yakovlev Yak-11 trainers.

YUGOSLAVIA

Yugoslav Air Force   *Jugoslovensko Ratno Vazduhoplovstvo*

Prior to 1918, Yugoslavia was a part of the Austro-Hungarian Empire, but the end of World War I brought independence. A number of Serb, Croat and Slovene officers who had flown with the Serbian Military Air Service, formed in 1912, were available as the nucleus of a Yugoslav Army Aviation Department in 1923. Initial equipment included Spad S-7C.1 and Dewoitine D-1C fighters, Breguet Br.19A.2 and 19B.2 reconnaissance-bombers, with a number of licence-built Brandenburg trainers. Although remaining part of the Army, a degree of independence was given to this air arm in 1930. In the meantime, Avia B.H.33 fighters, Potez XXV reconnaissance aircraft, Hanriot A.32 trainers and H.41 seaplanes (for the Yugoslav Navy) were obtained.

In 1935 the Yugoslav Army Air Corps was operating some forty-four squadrons with 440 aircraft. Modernization started with orders for Dewoitine D.500 and Hawker Fury fighters, Bristol Blenheim I, Dornier Do 17K and Savoia-Marchetti S.M.79 bombers, with some Dornier Do 22 general-purpose aircraft. At the time of the outbreak of World War II, the YAAC was receiving Hawker Hurricane, Curtiss P-40B Tomahawk and Messerschmitt Bf 109E fighters, and Bf 108 and Westland Lysander army co-operation aircraft. The country was occupied by German forces in April 1941, after a revolution had overthrown the monarch and a pro-Axis government.

A number of YAAC personnel managed to escape to the United Kingdom, where Yugoslav squadrons flying Supermarine Spitfire fighters and Martin Baltimore bombers served with the Royal Air Force. The Axis powers partitioned Yugoslavia, with Italy governing the Croat areas, forming a Croatian Air Force which flew Caproni Ca.310, Fiat G.50 and Messerschmitt Bf 109G fighters, Morane-Saulnier M.S.40 and Dornier Do 172 bombers, and a few AOP aircraft. There were some defectors from this force.

After the war ended in 1945, Yugoslavia became a republic largely, but not entirely, within the Russian sphere of influence, and the Jugoslovensko Ratno Vazduhoplovstvo (Yugoslav Air Force) was formed. Apart from the former RAF squadrons' equipment, the first aircraft came from Russia, and included Yakovlev Yak-3 and Yak-9 fighters, Ilyushin Il-2 and Il-10 ground-attack aircraft, Petlyakov Pe-2 bombers, Lisunov Li-2 (C-47) transports, and UT-2 and Po-2 trainers, so that by the late 1940s there were some four hundred aircraft. Some domestic aircraft production resumed, with the nationally-designed Aero 2 series of trainers, while a decline in Russian influence led the United States to provide aid in spite of Yugoslavia remaining a Communist country. Yugoslavia has in fact remained outside of the Warsaw Pact, to which every other East European Communist state belongs (with the exception of Albania). Equipment provided for the JRV during the early 1950s included 150 Republic F-47D Thunderbolt and 140 de Havilland Mosquito FB.6 fighter-bombers, plus a few Mosquito night-fighters. The nationally-designed S-49A fighter entered service but was soon withdrawn.

The JRV's first jets, Lockheed T-33A trainers, arrived in 1953, just preceding two hundred Republic F-84G Thunderjet fighter-bombers, Westland Dragonfly helicopters, and Douglas C-47 Dakota and Ilyushin Il-14 transports, with Aero 3 trainers. Canadair Sabre Mk. 2 and North American F-86D Sabres were obtained during the late 1950s. The 1960s saw F-86G Sabres, Mikoyan MiG-21Fs, and Soko Jastreb armed-trainers and Galeb jet trainers enter service.

The JRV still continues to receive its equipment from both sides of the Iron Curtain, supplementing these supplies with its own Aero trainers and Soko jet and armed-trainers.

Currently the JRV has 20,000 men and operates sixty Mikoyan MiG-21F interceptors; one hundred North American and Canadair F-86D/E Sabre fighters; thirty Lockheed RT-33A tactical reconnaissance aircraft; ninety Republic F-84G Thunderjets being replaced by 150 Soko Jastreb and thirty Kraguj ground-attack aircraft; fifteen Douglas C-47 Dakota, four DC-6Bs and six Ilyushin Il-14 transports; twenty Westland S-55 Whirlwind, fifteen Mil Mi-4, and two Sud Alouette III helicopters; 150 Soko

Galeb, seventy Lockheed T-33A, sixty Aero 3, and a few Mikoyan MiG-21UTI trainers. SA-2 Guideline surface-to-air missiles are deployed. A Sud Caravelle VIP transport is also operated.

ZAMBIA

Zambia Air Force

Formerly Northern Rhodesia, Zambia became independent in 1964, and immediately a small national air arm was formed using four Douglas C-47 Dakota and two Hunting Pembroke C.1 transports as initial equipment. Assistance was provided at first by the RAF, but more recently by an Italian company. Currently four hundred men strong, the ZAF operates two Douglas C-47 Dakota, four de Havilland Canada DHC-4 Caribou and six DHC-2 Beaver transports; five Agusta-Bell 205 helicopters; a few de Havilland Canada Chipmunk trainers; and eight Scottish Aviation Bulldog trainers. Eight Savoia-Marchetti SF.260 and two Soko Galeb trainers were delivered in 1971, with four Soko Jastreb armed-trainers.

Modern Aircraft Types

Interceptor   Fighter   Ground-Attack   Reconnaissance-Fighter

BAC LIGHTNING   Single-seat Mach 2·3 interceptor using two reheated Rolls-Royce Avon turbojets. F.1, F.2, F.3 and long-range F.6 for RAF, with T.4 and T.5 two-seat conversion trainers. Multi-mission variant, F.53, for Saudi Arabia and Kuwait, with T.55 trainers. Former English Electric design.

BAC SCIMITAR   Single-seat Mach 0·95 carrier-borne fighter using two Rolls-Royce Avon turbojets. F.1 version only. Remains in Fleet Air Arm service as tanker aircraft for in-flight refuelling. Former Supermarine design.

BAC-BREGUET JAGUAR   Single-seat Mach 1·5 strike-fighter using two reheated Rolls-Royce/Turboméca RB.172 turbofans. 164 'S' strike and 36 'B' two-seat conversion trainers for RAF; French Air Force equivalents 'A' and 'E'. 'M' is carrier-borne version for Aéronavale. Deliveries early 1970s.

CHANCE-VOUGHT, *see* Ling-Temco-Vought

CONVAIR, *see* General Dynamics

DASSAULT ETENDARD   Single-seat fighter, II for French Air Force with two Turboméca Gabizo turbojets as prototype only, but single SNECMA Atar turbojet IV-M in service with Aéronavale.

DASSAULT MIRAGE   Single-seat Mach 2·2 interceptor. IIIE using one reheated SNECMA Atar turbojet for French Air Force. Fighter-bomber variants: IIICJ – Israel; IIIEZ and CZ – South Africa; Swiss-built III-S; Australian-built III-O. Reconnaissance version IIIR, ground-attack version, 5, for Israel (suspended), Belgium, Libya, Iraq and Peru. Trainer, IIIB, with two seats. Mirage F.1 is interceptor for French Air Force without delta wing and using separate wing and tailplane. G is variable-geometry F.

DASSAULT SUPER MYSTÈRE   Single-seat Mach 1·2 interceptor using one reheated SNECMA Atar turbojet. B-2 in service French Air Force and Israel.

DE HAVILLAND, *see* Hawker Siddeley

DOUGLAS, *see* McDonnell Douglas

ENGLISH ELECTRIC, *see* BAC

FIAT G.91R Single-seat Mach 0·9 fighter-bomber with single Rolls-Royce Orpheus turbojet. 1, 1A and 1B – Italian Air Force; 3 and 4 – West Germany (including licence-built 3s); 4 – Portugal. G.91T, training version with two seats for Germany and Italy; G.91Y, Mach 0·95 version for Italian Air Force using two reheated General Electric J85 turbojets.

FOLLAND, *see* Hawker Siddeley

FAIRCHILD-HILLER F-84 THUNDERJET Single-seat subsonic fighter-bomber no longer in USAF service. Developments: swept-wing F-84F Thunderstreak and RF-84F Thunderflash (reconnaissance version) using one Wright J65 turbojet, still in service in a number of countries. Republic design.

FAIRCHILD-HILLER F-105 THUNDERCHIEF Single-seat Mach 2·2 interceptor and strike aircraft using one reheated Pratt and Whitney J75 turbojet. In service with USAF. Republic design.

GENERAL DYNAMICS F-102 DELTA DAGGER Single-seat Mach 1·25 interceptor using one reheated Pratt and Whitney J57 turbojet. F-102A for USAF, and some delivered second-hand to RHAF in 1970. F-106 Delta Dart is Mach 2·3 development using reheated J75 engine. TF-102 is two-seat conversion trainer. Convair design.

GENERAL DYNAMICS F-111 Twin-seat Mach 2·5 tactical fighter with variable-geometry wings and two reheated Pratt and Whitney TF30 turbofans. F-111A for USAF, F-111C on order for RAAF, RF-111 reconnaissance and FB-111 bomber versions for USAF. F-111B for USN and F-111K for RAF cancelled.

GRUMMAN A-6 INTRUDER Two-seat Mach 0·95 carrier-borne strike aircraft using two Pratt and Whitney J52 turbojets for USN and USMC. EA-6B is a four-seat electronics countermeasures version.

GRUMMAN F-14 TOMCAT Two-seat Mach 3 carrier-borne interceptor with variable-geometry and two reheated Pratt and Whitney TF30 turbofans. Deliveries to USN during 1970s.

HAWKER SIDDELEY GNAT Single-seat Mach 0·95 fighter for Finland and India (licence-built) and two-seat trainer for RAF, using one Rolls-Royce Orpheus turbojet. Folland design.

HAWKER SIDDELEY HARRIER Single-seat Mach 1·2 vertical take-off strike-fighter using one Rolls-Royce Pegasus turbojet. FGA.1 for RAF. More powerful version for USMC (AV-8A) was originally to be built under licence by McDonnell Douglas. T.2 is two-seat conversion trainer for RAF.

HAWKER SIDDELEY HUNTER Single-seat Mach 0·9 fighter-bomber using one Rolls-Royce Avon turbojet. Several marks, including F.2 and F.5 with Sapphire engines, for fighter-bomber, ground-attack, photographic and conversion training (two-seat) duties. In service with many air forces. Hawker design.

HAWKER SIDDELEY SEA HAWK Single-seat subsonic carrier-borne fighter-bomber using one Rolls-Royce Nene turbojet. Several versions, for Fleet Air Arm, Royal Netherlands, Federal German, and Indian Navies. Hawker design.

HAWKER SIDDELEY SEA VIXEN Two-seat Mach 1 carrier-borne twin-boom interceptor using two Rolls-Royce Avon turbojets. Production model F(AW).2 for Fleet Air Arm. De Havilland design.

HAWKER SIDDELEY VENOM Single-seat subsonic fighter-bomber using one de Havilland Ghost turbojet. Built under licence in several countries, and several marks produced, including carrier-borne Sea Venom for Fleet Air Arm and Royal Australian Navy. De Havilland design.

HAWKER SIDDELEY VAMPIRE Single-seat subsonic fighter-bomber using one de Havilland Goblin turbojet. Built under licence in several countries and in several versions for fighter, night-fighter (two-seat), fighter-bomber and trainer (two-seat basic trainers which are still to be found in service) duties, with carrier-borne variants. De Havilland design.

HELWAN HA-300 Single-seat lightweight fighter, production models of which are planned to have two E-300 turbojets and a speed of Mach 2. Egyptian project.

HINDUSTAN HF-24 MARUT Single-seat Mach 1 fighter with two Bristol Siddeley Orpheus turbojets. In service with Indian Air Force.

LTV F-8 CRUSADER Single-seat Mach 1·7 carrier-borne interceptor using one reheated Pratt and Whitney J57 turbojet. A few remain in service with the United States Navy, including the RF-8G reconnaissance version, and with the Aéronavale. Chance-Vought design.

LTV A-7 CORSAIR II Single-seat subsonic carrier-borne strike aircraft using one Pratt and Whitney TF30 turbofan or one Allison-built Rolls-Royce Spey (A-7E for USN). Also in service with USAF. Similar in appearance to Crusader.

LOCKHEED U-2 Single-seat Mach 0·8 high-altitude reconnaissance aircraft using one Pratt and Whitney J75. WU-2 is used for weather flights; U-2D has two seats. In service with USAF.

LOCKHEED SR-71 Two-seat Mach 3 interceptor using two reheated Pratt and Whitney J58 turbojets. Mainly used for reconnaissance as a U-2 replacement.

LOCKHEED F-104 STARFIGHTER Single-seat Mach 2·2 interceptor and strike-fighter using one reheated General Electric J79 turbojet. F-104A/B/C/D versions for USAF, with some for Nationalist China, Pakistan and Greece; F-104G European-built multi-mission development for NATO members; F-104J for Japan; CF-104 Canadian-built; some RF-104 reconnaissance variants.

McDONNELL DOUGLAS F-15 Single-seat Mach 3 interceptor using two reheated Pratt and Whitney JTF22 turbofans planned for USAF service in the mid-1970s.

McDONNELL DOUGLAS F-4 PHANTOM II Two-seat Mach 2·4 strike-fighter using two reheated General Electric J79 or Rolls-Royce Spey (F-4K and F-4M) turbofans. Reconnaissance, tanker and carrier-borne developments. In service with USAAF, USN, USMC, RAF, Fleet Air Arm, Iran, Israel, Japan, Korea and Federal Germany, with leased models operated by RAAF.

McDONNELL DOUGLAS A-1 SKYRAIDER Single-seat or two-seat counter-insurgency aircraft with a maximum speed of some 300 mph and a Wright radial engine. Used by French Air Force and South Vietnamese Air Force. Variants surviving are single-seat A-1H/J and two-seat A-1E.

McDONNELL DOUGLAS A-4 SKYHAWK Single-seat

Mach 0·9 carrier-borne attack aircraft with a single Pratt and Whitney J52 turbojet. A variety of models are in USN, USMC, Israeli, New Zealand and Royal Australian Navy service, along with the two-seat advanced trainer version, the TA-4.

McDONNELL DOUGLAS F-101 VOODOO Single-seat Mach 1·85 interceptor with two reheated Pratt and Whitney J57 turbojets. F-101B two-seat; RF-101A/C reconnaissance version; TF-101F conversion-trainer with two seats; CF-101B is Canadair-built version for Canadian Armed Forces. Other types in USAF service.

MIKOYAN MiG-15 'FAGOT' Single-seat subsonic fighter using one Klimov VK-1 turbojet. Used by all Communist bloc air forces at one time or another, and built in Poland and China (Shenyang F-2) as well as the USSR. MiG-15UTI 'Midget' is two-seat conversion and advanced trainer development.

MIKOYAN MiG-17 'FRESCO' Single-seat Mach 0·95 fighter relegated to ground-attack duties, using one Klimov VK-1A turbojet. Also built in Poland and China (Shenyang F-4), and used by all Communist bloc air forces. 'Fresco A/B/C' versions without reheat, but 'Fresco D' so equipped. No reheat 'Fresco E'.

MIKOYAN MiG-19 'FARMER' Single-seat Mach 1·3 fighter using two reheated Klimov RD-9F turbojets. Chinese-built versions (Shenyang F-6) supplied to Pakistan. Used by all Communist bloc air forces. A number of versions; 'Farmer D' all-weather interceptor development.

MIKOYAN MiG-21 'FISHBED' Single-seat Mach 2 interceptor using one reheated R37F turbojet. In service with most Communist bloc air forces, with some Chinese production (Shenyang F-8), but also supplied to Finland and India (some licence-production). A number of developments, of which the most important are the widely-used all-weather MiG-21FL 'Fishbed D', the Soviet Air Force's much-improved Mach 2·5 MiG-21PF 'Fishbed F', and the two-seat conversion trainer, MiG-21UTI 'Mongol'.

MIKOYAN MiG-23 'FOXBAT' Single-seat Mach 3 interceptor using two reheated turbojets. In production for Soviet Air Force.

MIKOYAN MiG- 'FLOGGER' Single-seat Mach 2·5 interceptor with variable-geometry and one reheated turbojet.

NORTH AMERICAN F-86 SABRE Single-seat Mach 0·9 fighter using one General Electric J47 in F-86A/E/F/H day-fighter and reheated J47 in F-86D all-weather interceptor versions. F-86K is Fiat-built F-86D. F-86L development of F-86D. Australian and Canadian-built versions by Commonwealth and Canadair respectively used Rolls-Royce Avon engines.

NORTH AMERICAN F-100 SUPER SABRE Single-seat Mach 1·3 interceptor and fighter-bomber using one reheated Pratt and Whitney J57 turbojet. In service with USAF, and delivered to France, Nationalist China, Denmark and Turkey.

NORTHROP F-5 FREEDOM FIGHTER Single-seat Mach 1·5 tactical fighter using two reheated General Electric J85 turbojets. In service with many air forces, in addition to the USAF. F-5B is two-seat version to accommodate either navigator/observer or pupil, CF-5 is Canadair-built version for CAF, NF-5 is also Canadian-built for Royal Norwegian Air Force. SF-5 is Spanish-built. T-38 Talon advanced trainer is closely related to F-5B, which it is similar to in appearance and powerplant.

PANAVIA 200 PANTHER Two-seat Mach 2·5 multi-role combat aircraft using variable-geometry and two reheated Rolls-Royce/MAN RB.199 turbofans. Deliveries during the mid-1970s to RAF (385), Luftwaffe (420), and Italian Air Force (100).

REPUBLIC, *see* Fairchild-Hiller

SAAB-JA29 TUNNAN Single-seat subsonic fighter using one licence-built de Havilland Ghost turbojet. A few remain on second-line duties in Sweden and Austria.

SAAB-JA32 LANSEN Single-seat Mach 0·9 ground-attack fighter using one reheated licence-built Rolls-Royce Avon turbojet. A few remain in RSwAf service.

SAAB-J35 DRAKEN Single-seat Mach 2 interceptor using one reheated licence-built Rolls-Royce Avon turbojet. In RSwAF and RDAF service, while Finland has just received twelve. J35X is a two-seat version for conversion training.

SAAB-J37 VIGGEN Single-seat Mach 2·2 multi-role aircraft using one reheated licence-built JT8D turbofan. AJ 37 is a strike fighter, JA 37 interceptor, S 37 reconnaissance, and Sk 37 two-seat conversion trainer.

SUKHOI Su-7 'FITTER'   Single-seat Mach 1·6 ground-attack fighter with one reheated TRD 31 turbojet. Su-7UTI 'Moujik' is a conversion trainer development. Used by many Communist bloc air forces.

SUKHOI Su-9 'FISHPOT'   Single-seat Mach 1·8 interceptor using Su-7 powerplant and fuselage. In limited use, mainly in Soviet Air Force, as a slower, but longer-range partner to MiG-21.

SUKHOI Su-11 'FLAGON'   Single-seat Mach 2·5 interceptor using two reheated TRD 31 turbojets. V/STOL development exists as prototype only, but standard version in Soviet service.

SUPERMARINE, *see* BAC

TUPOLEV Tu- 'FIDDLER'   Two-seat Mach 1·75 interceptor using two reheated turbojets. In limited service.

VFW-FOKKER Vak 191B   Single-seat VTOL reconnaissance-fighter using one Rolls-Royce RB.193 and two RB.162 turbofans. Planned as a Fiat G.91 replacement, this aircraft will not proceed beyond the prototype stage.

YAKOVLEV Yak-25 'FLASHLIGHT'   Two-seat Mach 0·95 all-weather fighter using two 37V turbojets. Mainly in Soviet service, but now obsolescent.

YAKOVLEV Yak-28 'FIREBAR'   Two-seat Mach 0·9 all-weather fighter using two R.37F turbojets. Obsolescent.

YAKOVLEV Yak 'MANDRAKE'   Single-seat Mach 0·8 high-altitude reconnaissance aircraft using two Klimov RD-9 turbojets. In Soviet service.

YAKOVLEV Yak 'FREEHAND'   Single-seat twin-turbofan V/STOL strike-fighter of Mach 1·2 performance.

Bomber   Reconnaissance-Bomber   Tanker

AVRO, *see* Hawker Siddeley

BAC CANBERRA   Two/three-seat Mach 0·8 medium bomber with two Rolls-Royce Avon or Sapphire jets. A number of versions, including night-fighter, reconnaissance, and trainer, are in service in a large number of air forces. Also built in Australia. American-built versions are the Martin B-57 and

RB-57, bomber and reconnaissance bombers respectively. Latest development is the conversion of refurbished aircraft to target-tugs. English Electric design.

BOEING B-47 STRATOJET Three/six-seat Mach 0·85 strategic and reconnaissance-bomber (RB-47H) using six General Electric J47 turbojets. Number of versions for weather-reconnaissance, RB-47K and WB-47; missile carrying, DB-47B; and photo-reconnaissance, RB-47E. USAF only, but now most of these aircraft are in reserve.

BOEING B-52 STRATOFORTRESS Six-crew Mach 0·85 strategic heavy bomber with eight Pratt and Whitney TF33 turbofans. Several versions, all for USAF, of which only the B-52G/H remain in service.

BOEING KC-135 STRATOTANKER, see C-135 transport

CONVAIR, see General Dynamics

DASSAULT MIRAGE IV Two-seat Mach 2·2 strategic bomber using two reheated SNECMA Atar turbojets. Service version is IV-A, and sole user is the French Air Force.

DOUGLAS, see McDonnell Douglas

ENGLISH ELECTRIC, see BAC

GENERAL DYNAMICS B-57F CANBERRA, see BAC

GENERAL DYNAMICS B-58 HUSTLER Three-seat Mach 2·1 strategic bomber using four reheated General Electric J79 turbojets. Convair design.

HANDLEY PAGE VICTOR Five-seat Mach 0·95 strategic bomber using four Rolls-Royce Conway (B.2) or Sapphire (B.1) turbojets. B.1s no longer in service and B.2s converted to tankers for inflight refuelling. RAF sole user.

HAWKER SIDDELEY BUCCANEER Two-seat Mach 1 carrier-borne low level strike-bomber, using two Rolls-Royce Spey turbofans (S.2) or Gyron Junior turbojets (S.1) S.1 operated by Fleet Air Arm; S.2 FAA and RAF, S.50 export version of S.2 for South Africa, equipped with rocket assistance. Blackburn design.

HAWKER SIDDELEY VULCAN Five-seat Mach 0·95 strategic bomber using four Rolls-Royce Olympus turbojets. Versions B.1 and B.2; latter remains in RAF service (only user). Avro design.

ILYUSHIN Il-28 'BEAGLE' Four-seat Mach 0·8 medium-bomber with two Klimov VK-1 turbojets. Training version is Il-28U 'Mascot'. In service with many Communist bloc air forces.

LOCKHEED KC-130 HERCULES, *see* transport

McDONNELL DOUGLAS B-66 DESTROYER Three-seat Mach 0·9 medium bomber with two Allison J71 turbojets. A development of the carrier-borne A-3 Skywarrior for USAF. Versions include RB-66A/B for reconnaissance, WB-66 for weather-reconnaissance. Most of these have been withdrawn from service, but EB-66B remains for electronics counter-measures. Douglas design.

MYASISHCHEV Mya-4 'BISON' Mach 0·8 strategic reconnaissance-bomber using four Mikulin AM-3D turbojets. Frequently operated on over-water missions.

NORTH AMERICAN ROCKWELL RA-5C VIGILANTE Two-seat Mach 2·1 carrier-borne attack aircraft using two reheated General Electric J79 turbojets. Original versions are A-5A/B/C, but mainly converted to RA-5C standard for reconnaissance duties.

TUPOLEV Tu-16 'BADGER' Mach 0·85 medium-range bomber using two Mikulin AM-3M turbojets. Frequently used on maritime-reconnaissance missions, although basically an aircraft for overland operations. In use with several Communist bloc air forces.

TUPOLEV Tu-20 'BEAR' Long-range turboprop bomber using four Kuznetsov NK-12M turboprops. Mainly with Soviet Air Force.

TUPOLEV Tu-22 'BLINDER' Long-range Mach 1·5 strike-bomber with two reheated tail-mounted turbojets. Replacing many Soviet 'Bears', 'Badgers' and 'Bisons'.

Maritime-Reconnaissance Anti-Submarine Amphibian Airborne Early Warning Flying-Boat

AVRO, *see* Hawker Siddeley

BERIEV BE-10 'MALLOW' Mach 0·8 maritime-reconnaissance flying-boat using two AL-7PB turbojets. Soviet Navy only.

BERIEV BE-12 'MAIL' Maritime-reconnaissance amphibian using two Ivchenko AI-20 turboprops. Soviet Navy only.

BREGUET Br. 1050 ALIZÉ Three-seat 300 mph carrier-borne anti-submarine aircraft using one Rolls-Royce Dart turboprop. In service with the French and Indian Navies.

BREGUET Br. 1150 ATLANTIC Maritime-reconnaissance aircraft using two licence-built Rolls-Royce Tyne turboprops. Max. speed 380 mph. In service with French, German, Dutch and Italian Navies.

CANADAIR CL-28 ARGUS, *see* Bristol Britannia transport.

FAIREY, *see* Westland

GRUMMAN S-2E TRACKER Four-seat carrier-borne 300 mph anti-submarine aircraft using two Wright R-1820 piston engines. E-1B Tracer is airborne-early-warning version. C-1A Trader is carrier onboard-delivery transport version. In service with USN, and numerous other navies.

GRUMMAN E-2A HAWKEYE Five-seat carrier-borne 300 mph anti-submarine aircraft using two Allison T56 turboprops. In service with USN only. C-2A Greyhound is carrier onboard-delivery transport version.

GRUMMAN HU-16 ALBATROSS Five-crew 240 mph maritime-reconnaissance, rescue and transport amphibian using two Wright R-1820 piston engines. Plans exist for turboprop conversion. Operated by numerous navies and air forces, and US Coastguard (HU-16E).

HAWKER SIDDELEY NIMROD MR.1 Mach 0·8 maritime-reconnaissance development of de Havilland Comet airliner, but uses four Rolls-Royce Spey turbofans. In service with RAF, and order for South African Air Force possible.

HAWKER SIDDELEY SHACKLETON Maritime-reconnaissance bomber using four Rolls-Royce Griffon piston engines. Being replaced by Nimrods. Latest version, MR.3, in service RAF and South African Air Force. Some airborne-early-warning versions. Avro design.

LOCKHEED P-2 NEPTUNE Maritime-reconnaissance bomber using two Wright R-3350 piston engines, sometimes aided by two Westinghouse J34 turbojets (P-2H). In service with a number of navies and air forces. Also built under licence in Japan by Kawasaki. Kawasaki P-2J is a General Electric T64 turboprop development of licence-built P-2Hs, with longer fuselage.

LOCKHEED P-3 ORION   Maritime-reconnaissance aircraft using four Allison T56 turboprops. Speed 400 mph and up to twelve crew. In service with USN, RAAF, RNZAF, CAF (licence-built).

LOCKHEED S-3   Mach 0·75 carrier-borne anti-submarine aircraft using two General Electric TF34 turbofans. In service with USN by mid-1970s.

LOCKHEED EC-121H   Long-range airborne-early-warning and electronic-reconnaissance development of C-121 transport (military version of Constellation airliner, now retired), powered by four Wright R-3350 piston engines. In service with USAF and USN.

SHIN MEIWA PS-1   Maritime-reconnaissance flying-boat using four licence-built General Electric T64 turboprops with maximum speed of 300 mph. In service with JMSDF.

TUPOLEV Tu-114 'MOSS'   Airborne-early-warning version of airliner, using four Kuznetsov NK-12MV turboprops. In Soviet use only.

Transport Communications Liaison

AVRO, *see* Hawker Siddeley

ANTONOV An-2 'COLT'   Light biplane transport using one ASH-621R radial engine: max. speed 150 mph. In service with most Communist bloc air forces, with production in several.

ANTONOV An-12 'CUB'   Medium transport using four Ivchenko AI-20 turboprops: max. speed 350 mph, 2,000 mile range, 44,000 lb. payload. Wide military and civil use.

ANTONOV An-14 'COLD'   Light STOL transport using two Ivchenko AI-14RF piston engines: max. speed 150 mph, 300 mile range, 2,000 lb. payload.

ANTONOV An-22 'COCK'   Heavy long-range transport using four Kusnetsov NK-12MA turboprops: max. speed 400 mph, 6,000 mile range, payload 175,000 lb.

ANTONOV An-24 'COKE'   Light transport with two Ivchenko AI-24T turboprops: max. speed 280 mph, 700 mile range,

11,000 lb. payload, or up to fifty passengers. Wide civil and military use.

ARMSTRONG-WHITWORTH, *see* Hawker Siddeley

BAC BRITANNIA Long-range transport with four Bristol Siddeley Proteus turboprops: max. speed 400 mph, range 5,000 miles, payload 33,000 lb. or 120 passengers. C.1 trooper, C.2 convertible passenger and freight. Canadair CC-106 Yukon licence-built version with stretched fuselage and Rolls-Royce Tyne turboprops, while Canadair CP-107 (CL-28) Argus maritime reconnaissance version with shortened fuselage, revised nose, and four Wright R-3350 radial engines. Short Belfast C.1 is a Tyne-engined development for heavy-lift duties: max. speed 350 mph, range 5,500 miles, payload 80,000 lb. or 250 troops. Bristol design.

BAC VC 10 C.1 Long-range transport with four tail-mounted Rolls-Royce Conway turbofans: max. speed 580 mph, range 3,500 miles, payload 60,000 lb. or 150 troops. RAF service only. Vickers design.

BEAGLE BASSET CC.1 Communications aircraft with two Rolls-Royce/Continental GIO-470A engines: max. speed 200 mph. Military version of Beagle 206.

BEECH U-21 Communications aircraft with two United Aircraft PT6A turboprops: max. speed 250 mph. Military version of Queen Air.

BEECH U-8 Communications aircraft with two Lycoming IGSO-480 engines: max. speed 240 mph. Military version of Twin Bonanza.

BEECH C-45 Communications and light transport aircraft using two Pratt and Whitney R-985 piston engines: max. speed 225 mph. Training versions, T-7 and T-11 Kansan. Military versions of Beech 18.

BOEING KC-135 STRATOTANKER Tanker version of C-135 transport, now largely replaced, with four Pratt and Whitney TF33 turbofans. Similar to Boeing 707 airliner. 560 mph, 4,000 mile range.

BREGUET 941S STOL transport using four Turboméca IIID-3 turboprops with max. speed 250 mph, 22,000 lb. payload, and range 1,500 miles. Prototypes still under test.

BRISTOL, *see* BAC

BRITTEN-NORMAN BN-2A ISLANDER Twin-engined light transport and communications aircraft using two Lycoming O-540-E4C5 piston engines: max. speed 160 mph, range 750 miles, nine passengers. Armed version Defender; Trislander three-engined stretched development.

CANADAIR CC-106 YUKON, see BAC Britannia

CASA C.207 AZOR Light transport with two Bristol Hercules radial engines: max. speed 250 mph, range 1,200 miles, 6,600 lb. payload, or thirty passengers.

CASA C.212 AVICAR STOL transport with two Garrett TPE turboprops: max. speed 225 mph, range 1,200 miles, payload 4,400 lb. or eighteen passengers.

CONVAIR, see General Dynamics

CTA C-95 BANDEREIRANTE Light transport and communications aircraft using two United PT6A turboprops: max. speed 280 mph, range 1,000 miles, nine passengers. Designed for Brazilian Air Force.

DASSAULT FALCON Light transport and communications aircraft using two General Electric CF700 turbofans: max. speed 520 mph, range 2,000 miles, eight passengers.

DASSAULT MD.315 FLAMANT Light transport with two SNECMA 12S piston engines: max. speed 240 mph, range 750 miles, ten passengers. Built for French armed services, but some sold to African countries.

DE HAVILLAND, see Hawker Siddeley

DE HAVILLAND CANADA DHC-2 BEAVER Light transport with single Pratt and Whitney R-985 radial engine: max. speed 150 mph. In world-wide service.

DE HAVILLAND CANADA DHC-3 OTTER Light transport with single Pratt and Whitney R-1340 radial engine: 2,000 lb. payload. In world-wide service.

DE HAVILLAND CANADA DHC-4 CARIBOU STOL transport with two Pratt and Whitney R-2000 radial engines: max. speed 250 mph, range 1,200 miles, payload 8,500 lb. or thirty troops. In world-wide service. USAF C-7.

DE HAVILLAND CANADA DHC-5 BUFFALO STOL transport with two General Electric T-64 turboprops: max. speed 280 mph, range 2,000 miles, payload 14,000 lb. or forty troops. In service in Canada, Brazil, US Army (C-8A).

DE HAVILLAND CANADA DHC-6 TWIN OTTER STOL light transport with two united PT6A turboprops: max. speed 200 mph, range 600 miles, payload 5,000 lb. or fifteen passengers. In service in a number of air forces and air arms.

DINFIA IA 35 HUANQUERO Light transport with two IAR 19C radial engines: max. speed 220 mph, six/eight passengers. But usually used as trainer or ambulance. Argentinian Air Force.

DINFIA IA 50 GUARANI II Light transport with two Turboméca Bastan VI-A turboprops: max. speed 300 mph, range 1,100 miles, payload 3,300 lb. or ten/fifteen passengers. Argentinian Air Force.

DORNIER Do 27 Liaison aircraft with single Lycoming GO-480 piston engine: max. speed 150 mph, five seats. In service in a number of countries.

DORNIER Do 28 SKYSERVANT Light transport and communications aircraft with two Lycoming IGSO-540 piston engines: max. speed 180 mph, range 1,000 miles, twelve seats. In service with German armed forces and a few other military customers.

DOUGLAS, *see* McDonnell Douglas

FAIRCHILD HILLER C-119K PACKET Medium transport of twin-boom layout with two Pratt and Whitney R-3350 radial engines. Relatively few survive in service, and most of these have been modified to C-119K standard with the addition of two General Electric J85 auxiliary turbojets and other equipment. AC-119G/K are gunship versions for use in South Vietnam.

FAIRCHILD HILLER C-123 PROVIDER Transport with two Pratt and Whitney R-2800 radial engines: max. speed 220 mph, range 1,500 miles, payload 22,000 lb. C-123K has been modified in the same way as the C-119K. In service with the USAF and a number of other air forces.

FIAT G.222 STOL transport with two General Electric T64 turboprops: max. speed 250 mph, range 2,500 miles, payload 20,000 lb. or forty passengers. In prototype stage, but production likely for Italian Air Force.

FOKKER F-27 Mk. 400M TROOPSHIP Transport with two Rolls-Royce Dart turboprops: max. speed 300 mph, range 1,100 miles, payload 12,500 lb. or forty-five troops. In use in Netherlands, Sudan, Argentine, Philippines, Indonesia. Licence-built by Fairchild Hiller in US.

GENERAL DYNAMICS C-131 Transport with two Pratt and Whitney R-2800 radial engines: max. speed 280 mph, range 1,400 miles, payload 18,000 lb. In service with USAF, CAF (licence-built) and one or two other air forces. Canadian versions have Napier Eland turboprops. EC-131G is a USAF electronics reconnaissance type. T-29 is navigational trainer. Convair design – military version of 440 airliner.

GRUMMAN C-1A TRADER, see Tracker ASW aircraft.

GRUMMAN C-2A GREYHOUND, see Hawkeye AEW aircraft.

HANDLEY PAGE HERALD 400 Transport with two Rolls-Royce Dart turboprops: max. speed 280 mph, range 700 miles, fifty troops. In service with Royal Malaysian Air Force.

HAWKER SIDDELEY ANDOVER C.1 Transport with two Rolls-Royce Dart turboprops: max. speed 280 mph, range 1,100 miles, payload 16,000 lb. or sixty troops. In service with RAF, while civil version, HS 748, is, with some modifications, in service with a number of air forces. Avro design.

HAWKER SIDDELEY ARGOSY C-1 Transport of twin-boom layout, with four Rolls-Royce Dart turboprops: max. speed 300 mph, range 2,000 miles, payload 22,000 lb. or eighty troops. RAF service only. Armstrong-Whitworth design.

HAWKER SIDDELEY COMET C.4. Long-range transport with four Rolls-Royce Avon turbojets: max. speed 560 mph, range 3,000 miles, 100 passengers. In RAF service only, although RCAF operated two C.1As at one time. De Havilland design.

HAWKER SIDDELEY DEVON C.1. Light transport and communications aircraft with two Rolls-Royce Gipsy Queen engines: max. speed 240 mph, range 1,000 miles, eight passengers. In service with numerous air forces for VIP transport duties. Sea Devon is version for Royal Navy. Military version of Dove. De Havilland design.

HAWKER SIDDELEY HERON C.1 Basically a four-engined Devon or Dove, with a longer fuselage, seating fourteen/seventeen passengers.

HELIO U-10 SUPER COURIER STOL utility aircraft with

single Lycoming GU-480 piston engine: max. speed 150 mph, 150 mph, range 600 miles, five seats.

ILYUSHIN Il-14 'CRATE' Transport with two Shvertsov ASH-82T piston engines: speed 260 mph, range 1,000 miles, thirty-two passengers. A development of the Ilyushin Il-12 'Coach'. In service with many Communist bloc air forces and the Indian Air Force.

ILYUSHIN Il-18 'COOT' Medium-range transport with four Ivchenko AI-20 turboprops: max. speed 400 mph, range 3,000 miles, payload 30,000 lb. or 120 passengers. In service with several Communist bloc air forces.

LOCKHEED C-130 HERCULES Transport aircraft with four Allison T56 turboprops: max. speed 400 mph, range 4,000 miles, payload 45,000 lb. Numerous versions: transports are C-130A/B/E, with K as the RAF version of the E, D-130C was STOL prototype, C-130D ski-equipped, RC-130B survey, HC-130P tanker, HC-130G US Coastguard, HC-130H rescue, AC-130 close-support gunship for use in South Vietnam. Widely used.

LOCKHEED C-140 JETSTAR Transport and electronics aircraft using four Pratt and Whitney J-60-P-5 turbojets: max. speed 560 mph, range 2,000 miles, twenty seats.

LOCKHEED C-141 STARLIFTER Heavy transport with four Pratt and Whitney TF33 turbofans: max. speed 550 mph, range 5,500 miles, payload 88,000 lb. With USAF only.

LOCKHEED C-5 GALAXY Heavy transport with four General Electric TF39 turbofans: max. speed 550 mph, range 5,600 miles, payload 250,000 lb. with USAF only.

MAX HOLSTE, *see* Rheims

McDONNELL DOUGLAS C-9A NIGHTINGALE Transport, mainly ambulance duties, with two Pratt and Whitney JT8D turbofans; max. speed 560 mph, range 1,500 miles, forty stretchers. With USAF only, but military version of DC-9 transport.

McDONNELL DOUGLAS C-133B CARGOMASTER Heavy transport with four Pratt and Whitney T34 turboprops: max. speed 300 mph, range 3,000 miles, 100,000 lb. With USAF only.

McDONNELL DOUGLAS C-118 LIFTMASTER Transport with four Pratt and Whitney R-2800 radial engines: max. speed 280 mph, range 2,500 miles, sixty troops. Military version of DC-6. Still fairly widely used.

McDONNELL DOUGLAS C-54 SKYMASTER Transport with four Pratt and Whitney R-2000 radial engines: max. speed 250 mph, range 2,500 miles, fifty troops. Military DC-4.

McDONNELL DOUGLAS C-47 DAKOTA Transport with two Wright R-1820 radial engines: max. speed 200 mph, range 1,200 miles, thirty troops. C-117 Skytrain is modernized version for US Navy. AC-47 is close-support gunship version for use in South Vietnam. Military DC-3. USSR-built version is Lisunov Li-2 'Cab'.

MESSERSCHMITT-BOLKOW-BLOHM HFB HANSA Light transport and communications aircraft with two General Electric CJ610 turbojets: max. speed 520 mph, range 1,300 miles, eight seats. In Luftwaffe service.

MITSUBISHI Mu-2 Light transport, liaison and reconnaissance aircraft with two AiResearch TPE-331 turboprops: max. speed 300 mph, range 1,200 miles, seven seats. In service with JGSDF and JASDF.

NAMC YS-11 Transport with two Rolls-Royce Dart turboprops: max. speed 300 mph, range 1,200 miles, payload 16,000 lb. In service with JASDF and JMSDF.

NAMC XC-1 STOL transport with two Pratt and Whitney JT8D turbofans: max. speed 450 mph, range 1,800 miles, payload 18,000 lb. In JASDF service by mid-1970s.

NORD 2501 NORATLAS Transport of twin-boom layout with two licence-built Rolls-Royce Hercules radial engines: max. speed 200 mph, range 1,000 miles, 25,000 lb. payload. Only a few remain in service.

PIAGGIO P.166M Light tactical transport with two pusher-propeller IGSO-540 piston engines: max. speed 250 mph, range 1,300 miles, ten passengers. With Italian and South African armed forces.

PIAGGIO-DOUGLAS PD-808 Light transport, navigational trainer, and airways navigational equipment check aircraft, with two Rolls-Royce Viper turbojets: max. speed 500 mph, range 1,200 miles, six/ten seats. In service with Italian Air Force.

PILATUS PC-6 SUPER PORTER Single United PT6A-20 turboprop-powered utility and liaison aircraft: max. speed 150 mph, range 500 miles. In use by Swiss Air Force and Australian Army.

RHEIMS AVIATION MAX HOLSTE 1521M BROUSSARD Light transport with single Pratt and Whitney R-85 radial engine: max. speed 160 mph, range 500 miles. In service with French Army and numerous African air arms.

SCOTTISH AVIATION TWIN PIONEER STOL transport with two Alvis Leonides radial engines: max. speed 165 mph, range 750 miles, twelve/sixteen seats. In RAF and RMAF service.

SHORT BELFAST C.1, *see* BAC Britannia

SHORT 3M SKYVAN Light transport with two Garrett TPE-331 turboprops: max. speed 220 mph, range 600 miles, payload 5,000 lb. or twenty-two passengers. Entering service with a small number of air forces.

TRANSALL C.160 Transport with two licence-built Rolls-Royce Tyne turboprops: max. speed 300 mph, range 2,500 miles, payload 35,000 lb. In French, German and South African service.

Helicopters

AGUSTA A 106 Single-seat ASW helicopter for operation from warships. Turboméca-Agusta TAA 230 turbine engine, max. speed 100 mph.

BELL 47 SIOUX Two/five-seat helicopter for training, communications, AOP and liaison duties. Lycoming VO-435 engine, max. speed 100 mph. OH-13, US Army; TH-13, US Navy. Licence-production by Westland, Agusta, and Kawasaki. 47J is cabin version with five seats.

BELL 204/205 IROQUOIS Utility helicopter with up to fifteen seats. Lycoming T53 turbine engine in standard version, but General Electric, United, Rolls-Royce and Turboméca powerplants also used. HU-1A US Army, designation for first of series. Licence-production by Dornier, Agusta and Kawasaki.

BELL 206A KIOWA Utility helicopter with eight seats. Allison T63 turbine engine, max. speed 140 mph. TH-57A Sea Ranger US Navy; OH-58A Kiowa US Army. Civil version JetRanger.

BELL 209 AH-1 HUEYCOBRA Gunship version of Iroquois with same Lycoming engine.

BOEING-VERTOL 107 SEA KNIGHT Twin-rotor transport helicopter with twenty-five seats. Two General Electric CT58 turbines, max. speed 160 mph. CH-46 US Navy; UH-46 US Army.

BOEING-VERTOL 114 CHINOOK Twin-rotor transport helicopter with forty-five seats. Two Lycoming T55 turbines, max. speed 180 mph. US Army CH-47.

HUGHES 300 Three-seat development of 269A two-seat training helicopter. Lycoming H10 engine, max. speed 80 mph. US Army TH-55A.

HUGHES 500 CAYUSE Utility, AOP and liaison helicopter with six seats. Allison T63 turbine engine, max. speed 150 mph. US Army OH-6A.

KAMAN H-43 HUSKIE Five-seat contra-rotating rotor utility helicopter, mainly used for rescue, photography or fire fighting. US Navy OH-43D Pratt and Whitney R-1340 piston engine; USAF HH-43B Lycoming T53 turbine, max. speed 120 mph.

KAMAN UH-2 SEASPRITE Utility helicopter used mainly for rescue, but with twelve seats. General Electric T58 turbine engine, max. speed 160 mph. US Navy.

KAMOV Ka-25 'HORMONE' Naval helicopter with contra-rotating rotors, mainly equipped for ASW. Two Glushenkov turbine engines, max. speed 130 mph.

KAWASAKI KH-4, *variant of* Bell 47

MIL Mi-4 'HOUND' Utility and transport helicopter with fourteen seats. ASH-82V piston engine, max. speed 130 mph. In extensive use, particularly in Communist bloc.

MIL Mi-6 'HOOK' Heavy transport helicopter with 120 seats. Two Soloviev D-25 turbine engines, max. speed 200 mph.

MIL Mi-8 'HIP' Utility and transport helicopter with twenty seats. Two Isotov TB-2 turbines, max. speed 150 mph.

MIL Mi-10 'HARKE' Crane version of Mi-6 with shallow fuselage and twenty-eight seats.

SIKORSKY H-19 CHICKASAW Utility and transport helicopter with ten seats. Wright R-1300 piston engine, max. speed 110 mph. US Army UH-19D; USAF UH-19B; USMC CH-19E; USN UH-19F. Manufactured under licence by Mitsubishi and Westland. Westland S-55 Whirlwind used Alvis Leonides Major piston engine or Rolls-Royce Gnome turbine. S-55 civil version. In widespread use.

SIKORSKY H-34 CHOCTAW Utility and transport helicopter with twelve seats. Wright R-1820 piston engine, max. speed 130 mph. US Army CH-34A/C Choctaw; USN SH-34G/H/J Seabat; USMC UH-34D/E Seahorse. Licence-built by Sud and Westland. Westland S-58 Wessex used two Rolls-Royce Gnome turbines. S-58 civil version. In widespread use.

SIKORSKY H-3 SEA KING Transport (twenty-eight seats) and ASW helicopter. Two General Electric T58 turbines, max. speed 160 mph. SH-3A/D USN ASW and transport, RH-3A minesweeping. Licence-built by Agusta and Westland. Westland SH-3D Sea King uses two Rolls-Royce Gnome turbines. S-61 civil version.

SIKORSKY S-62 Utility helicopter with twelve seats. Single General Electric CT58 turbine, max. speed 100 mph. US Navy HH-52.

SIKORSKY CH-54A SKYCRANE Lift helicopter. Single Pratt and Whitney T73 turbine, max. speed 120 mph. US Army. Civil version S-64A.

SIKORSKY H-53 SEA STALLION Medium transport helicopter with forty seats. Two General Electric T64 turbines, max. speed 200 mph. USAF HH-53; USMC CH-53A. Civil version S-65.

SUD ALOUETTE II Light liaison helicopter, five seats. Single Turboméca Astazou turbine, max. speed 100 mph. In widespread use.

SUD ALOUETTE III Utility helicopter, seven seats. Single Turboméca Artouste turbine engine, max. speed 130 mph. In widespread use.

SUD SA.321 SUPER FRELON Heavy transport helicopter, thirty seats. Three Turboméca Turmo IIIC3 turbines, max. speed 160 mph. Aéronavale and Israel.

SUD SA.330 PUMA Transport helicopter, twelve seats. Two

Turboméca Turmo turbines, max. speed 170 mph. Production shared with Westland. In increasingly widespread use: British, French and other armed services.

SUD SA.341 GAZELLE   AOP and liaison helicopter. Single Turboméca Astazou turbine, max. speed 150 mph. With British and French Armies. Production shared with Westland.

WESTLAND S-55 WHIRLWIND: development of Sikorsky H-19.

WESTLAND S-58 WESSEX: development of Sikorsky H-34.

WESTLAND SH-3D SEA KING: development of Sikorsky H-3.

WESTLAND SIOUX: licence-built Bell 47.

WESTLAND SCOUT   Liaison, attack and AOP helicopter with four seats. Rolls-Royce Nimbus turbine, 120 mph. Naval version is Wasp, primarily a ship-borne ASW helicopter. With British Army and Fleet Air Arm, and some foreign countries.

WESTLAND WG.13 LYNX   Utility and ASW helicopter, with fourteen seats. Two Rolls-Royce 360 turbines, max. speed 180 mph. In British and French services.

Trainer Armed-Trainer Tactical Reconnaissance Gunship Counter-Insurgency Operation

All aircraft are two-seat types, unless stated otherwise.

AERMACCHI MB.326   Basic jet trainer with single Rolls-Royce Viper turbojet, max. speed 500 mph. Basic model Italian Air Force, MB.326B Tunisia, MB.326F Ghana, MB.326 more powerful armed version, MB.326K licence-built armed version South Africa, MB.326H licence-built unarmed version Australia. Single-seat version available.

AERO 3   Basic trainer with single Lycoming O-453 piston engine. Yugoslav Air Force.

AERO L-29 DELFIN   Basic jet trainer with single M.701C turbojet, max. speed 500 mph. In service with many Communist bloc air forces, Nigeria and Uganda.

BAC JET PROVOST   Basic jet trainer with single Rolls-Royce Viper turbojet, max. speed 400 mph. BAC 145, pressurized version, with 167 Strikemaster as armed-trainer development of 145. In service with a number of air forces. Hunting design.

BAC LIGHTNING, conversion trainer, *see* interceptors
BAC CANBERRA, conversion trainer, *see* bombers
BEECH T-34 MENTOR Basic trainer with one Continental O-470 piston engine, max. speed 180 mph. In service with a number of air forces and navies, with Fuji KM-2 as four-seat cabin development.
BEECH T-42A BARON Navigational trainer with two Continental IO-470 piston engines, max. speed 220 mph. In US Army service.
BEECH T-11 KANSAN: navigational trainer development of C-45 transport.
CANADAIR CT-114 TUTOR Basic trainer with single General Electric J85 turbojet, max. speed 440 mph. In service with CAF, Malaya (armed).
CESSNA T-41 Basic trainer with single Continental O-300 (T-41A) or O-360 (T-41B), max. speed 140 mph. In service with a number of air forces and armies.
CESSNA T-37 Basic trainer with two licence-built Turboméca Maboré turbojets, max. speed 480 mph. T-37B trainer, A-37A attack conversion, A-37B production attack version. In USAF service.
DASSAULT MIRAGE IIIB, conversion trainer, *see* fighters
DASSAULT-BREGUET-DORNIER ALPHA JET Advanced trainer with two SNECMA Larzac turbofans, Mach 1. Service by mid-1970s for French and German Air Forces.
FIAT G.91T/1, conversion trainer, *see* fighters
FOLLAND, *see* Hawker Siddeley
FUJI KM-2, *see* Beech T-34 Mentor
FUJI T.1 Basic trainer with single Ishikawajima-Harima J3-IHI-3 turbojet, max. speed 480 mph. In JASDF service.
GRUMMAN TC-4C Navigational trainer with two Rolls-Royce Dart turboprops, max. speed 300 mph. In USN service.
GRUMMAN OV-1 MOHAWK Counter-insurgency aircraft (COIN) with two Lycoming T53-L turboprops, max. speed 320 mph. Reconnaissance versions. With US Army.
HAWKER SIDDELEY DOMINIE T.1 Navigational trainer with two Rolls-Royce Viper turbojets, max. speed 500 mph. Variant of HS 125 executive jet. De Havilland design.
HAWKER SIDDELEY GNAT T.1, *see* fighters

HAWKER SIDDELEY HUNTER T.66, conversion trainer, *see* fighters
HAWKER SIDDELEY HARRIER T.2, conversion trainer, *see* fighters
HAWKER SIDDELEY VAMPIRE T.55, *see* fighters
HINDUSTAN HT-2 Basic trainer with single Cirrus Major engine, max. speed 130 mph. Indian Air Force.
HINDUSTAN HJT-16 KIRAN Basic trainer with single Rolls-Royce Viper turbojet, max. speed 420 mph. Indian Air Force.
HISPANO HA-200E SUPER SAETA Basic trainer with two Turboméca Marboré turbojets, max. speed 400 mph. Spanish and Egyptian (licence-built) air forces.
HUNTING, *see* BAC
ILYUSHIN Il-28U, *see* bombers
LOCKHEED TF-104 STARFIGHTER, conversion trainer, *see* fighters
LOCKHEED T-33A Advanced trainer with single Allison J33 turbojet, max. speed Mach 0·8. Conversion from F-80C Shooting Star fighter. In world-wide service.
LOCKHEED AC-130 HERCULES: gunship conversion of C-130 transport.
McDONNELL DOUGLAS TA-4 SKYHAWK: advanced trainer variant of fighter.
McDONNELL DOUGLAS AC-47: gunship conversion of C-47 Dakota transport.
MITSUBISHI XT-2 Advanced trainer with two reheat Rolls-Royce RB.172 turbofans, max. speed Mach 1·5. Due for JASDF service by mid-1970s.
MIKOYAN MiG-15UTI 'MIDGET', *see* fighters
MIKOYAN MiG-21UTI 'MONGOL', *see* fighters
NORD 262C Navigational trainers with two Turboméca Bastan turboprops, max. speed 250 mph. With French armed services.
NORTH AMERICAN T-6 HARVARD Basic trainer with single Pratt and Whitney R-1340 radial engine, max. speed 200 mph. T-6G Texan up-rated version.
NORTH AMERICAN T-28 TROJAN Trainer with single Wright R-1820 radial engine, max. speed 280 mph. Numerous variants, often armed. In world-wide service.

NORTH AMERICAN T-2B BUCKEYE Trainer with two Pratt and Whitney J60 turbojets, max. speed 520 mph. In USN service.

NORTH AMERICAN T-39 SABRELINER Navigational trainer with two Pratt and Whitney J60 turbojets, max. speed 560 mph. With US forces.

NORTH AMERICAN OV-10A BRONCO COIN aircraft with two Garrett T76 turboprops, max speed 280 mph. Twin-boom structure. With USAF, USMC, Luftwaffe (target tug).

NORTHROP F-5B, *see* fighters

NORTHROP T-38 TALON, *see* F-5 fighter

PIAGGIO P.149D Basic trainer with single Lycoming GO-480 piston engine, max. speed 180 mph. In service with a number of air forces.

POTEZ CM170 SUPER MAGISTER Basic trainer with two Turboméca Marboré turbojets, max. speed 440 mph. In fairly widespread use.

SAAB-105XT Trainer with two General Electric CJ 610 turbojets, max. speed 560 mph. 105XT armed export version of Sk 60.

SAAB-91 SAFIR Basic trainer with single Lycoming O-360 piston engine, max. speed 180 mph. Current model 91D. In Swedish, Ethiopian, Austrian, Finnish and Tunisian service.

SCOTTISH AVIATION BULLDOG Basic trainer with single Lycoming IO-360 piston engine, max. speed 160 mph. Orders RSwAF, Kenya, option Swedish Army. Beagle design (prior to liquidation).

SOKO G-2A GALEB Basic trainer with single Rolls-Royce Viper turbojet, max. speed 500 mph. In Yugoslav service. J-1 Jastreb armed-trainer version, in Yugoslav service.

SUKHOI Su-7UTI, conversion trainer, *see* fighters

YAKOVLEV Yak-18 'MAX' Basic trainer with single Ivchenko AI-14RF radial engine, max. speed 200 mph. In Communist bloc service.

Air Observation Post

AERFER-AERMACCHI AM.3 STOL AOP and liaison aircraft with single Continental GTSIO-520-C piston engine:

max. speed 180 mph, range 500 miles, three seats. In Italian Army service.

CESSNA O-1 BIRD DOG AOP and liaison aircraft with single Continental O-470-11 piston engine: max. speed 120 mph. In widespread use.

CESSNA O-2A SUPER SKYMASTER Forward air control aircraft with two (one conventional, one with pusher-propeller) tandem-mounted Continental IO-360 piston engines in a twin-boom airframe: max. speed 240 mph, range 900 miles, six seats. O-2B is loudspeaker-equipped version. In USAF service.

NEIVA REGENTE Liaison (C-42) and AOP (L-42) aircraft with single Lycoming O-360 piston engine: max. speed 140 mph, range 500 miles, four seats. With Brazilian armed forces only.

YAKOVLEV Yak-12 'CREEK' Liaison aircraft with one Ivchenko AI-14R radial engine: max. speed 120 mph, four seats. In widespread Communist bloc service.

Captions to the Plates

front of jacket

One of the few flying-boats left, and one of the even rarer modern flying-boats, is the Shin Meiwa PS-1 anti-submarine turboprop flying-boat used by the Japanese Maritime Self-Defence Force. An initial fourteen will be in service by 1973. The example shown has obviously just taken off. (Photo: Japanese Self-Defence Agency.)

back of jacket

NATO exercises are an irresistible lure for Soviet reconnaissance aircraft, usually of the Independent Navy Air Fleet. Shown here is a Tupolev Tu-16 'Badger' reconnaissance-bomber which is being shepherded away by a McDonnell Douglas F-4K Phantom II of the Fleet Air Arm. The photograph was taken from the flight deck of the aircraft carrier HMS *Ark Royal*, from which the Phantom was operated. A Westland Gannet AEW.3 airborne-early-warning aircraft is to the left on the flight deck, and to the right is a Hawker Siddeley Buccaneer S.2 strike-bomber. The exercise was 'Northern Wedding' held in the autumn of 1970. (Photo: Central Office of Information.)

1 One of the world's most successful jet bombers, as well as one of the first, was the BAC Canberra, refurbished versions of which are finding a ready market. The aircraft shown is one of a number of B.62s recently delivered to the Fuerza Aérea Argentina. The photograph was taken during flight-testing prior to delivery, hence the British registration number just aft of the serial number on the fuselage. (Photo: BAC.)

2 De Havilland Canada DHC-6 Twin Otters are to be found in service with many of the world's armies, navies and air forces, including all three Argentinian services, as landplanes, seaplanes or with ski equipment fitted. This Armada Argentina ski-Twin Otter is operating over the southern part of the country. (Photo: DHC.)

3 One of the Royal Australian Air Force's four Dassault Mirage III-O fighter-bomber squadrons, on the ground while the aircraft receive attention. (Photo: RAAF.)

4 The Westland Wessex is a British-built and British-developed version of the Sikorsky S-58 helicopter. The example illustrated here is one of twenty-seven used by the Royal Australian Navy on anti-submarine duties. The bell-shaped protrusion on the underside is the housing for the sonar buoy, which the helicopter 'dunks' while the sonar operator on board listens for submarines. (Photo: RAN.)

5 Australia's Fleet Air Arm again! One of the fourteen Grumman S-2E Tracker anti-submarine aircraft and two of the ten Douglas A-4G Skyhawk fighter-bombers which are operated from the aircraft carrier HMAS *Melbourne*. (Photo: RAN.)

6 A remarkably versatile aircraft, this Austrian Air Force SAAB 105E is used for ground-attack duties, but the aircraft can also be used for training and liaison. (Photo: Austrian Air Force.)

7 Seen against a typical background is this Sud Alouette III utility helicopter of the Austrian Air Force. The Alouette III is one of the most successful helicopters to have been built outside of the USA or USSR. (Photo: Austrian Air Force.)

8 One of the two Sikorsky S-65S helicopters recently delivered to the Austrian Air Force by Sikorsky. This is one of Sikorsky's largest helicopters, and is operated by the USAF as the HH-53C. (Photo: Sikorsky.)

9 The Force Aérienne Belge has received eighty-eight Dassault Mirage 5 ground-attack fighters. This aircraft is a ground-attack development of the Mirage III series, and is on order for Libya and Peru. The maximum speed of Mach 2·2 is very high for an aircraft intended primarily for ground-attack duties. (Photo: Dassault.)

10 The Force Terrestre Belge (Belgian Army) operates a number of light aircraft and helicopters, including the Dornier Do 27 liaison aircraft seen here waiting to take off, and the Sud Alouette II helicopter. (Photo: Force Terrestre Belge.)

11 One of the most durable aircraft of the post-war period is the Grumman HU-16 Albatross amphibian, which was in production during the mid and late 1940s and remains in service with a large number of air forces today, mainly on rescue duties. One such example is this one belonging to the Canadian Armed Forces. (Photo: Canadian Armed Forces.)

12 A de Havilland Canada DHC-5 Buffalo tactical transport of the Canadian Armed Forces lands on a rough strip, typical of the terrain which this aircraft is designed to operate from. The Buffalo is in service at present with the USAF, Canadian Armed Forces, and Brazilian Air Force. (Photo: Hawker Siddeley.)

13 One of the most famous aircraft of all time, and the most successful transport ever, the Douglas C-47 Dakota is still far from being a rarity some thirty-six years after the first flights of its civil cousin, the DC-3. The example shown here belongs to the Fuerza Aérea Colombiana, but C-47s are in service with a number of African and Latin American air forces, some Eastern air forces, and the USAF and USN, which also operate the AC-47 gunship version and the modernized C-117 Skytrain. (Photo: Fuerza Aérea Colombiana.)

14 Still popular fighter aircraft, the North American F-86 Sabres first appeared around the time of the Korean War. A squadron of F-86Fs of the Fuerza Aérea Colombiana is shown here getting ready for flight. A Douglas C-54 Skymaster can be seen in the background. (Photo: Fuerza Aérea Colombiana.)

15 Seen flying at a great height, two developments of the highly successful Dassault Mirage series. Nearest to the camera is a Mirage F.1 interceptor, and further away is a Mirage G variable-geometry development of the F.1. The wings of the Mirage G are in the fully-swept position. Both aircraft are operated by the Armée de l'Air. (Photo: Dassault.)

16 A Dassault Mirage IV nuclear bomber takes off using rocket assistance. The Mirage IV is a twin-engined development of the standard Mirage, and sixty-three are operated in nine Armée de l'Air squadrons as part of France's *force de frappe*, or nuclear deterrent. (Photo: Dassault.)

17 A Rheims Aviation Max Holste 1521M Broussard (Bushranger) of the Armée de la Terre (French Army). These aircraft are comparable to the de Havilland Canada Beaver and are used for light transport and communications duties. France has supplied Broussards to the air arms of nearly every former French African colony. (Photo: Armée de la Terre.)

18 The Sud SA.330 Puma is a helicopter finding acceptance with a steadily growing number of air forces and air arms. Produced jointly by Sud and Westland, users include the French and British armed services, and the South African Air Force. The example shown here, about to land, belongs to the Armée de la Terre. (Photo: Armée de la Terre.)

19 Three McDonnell Douglas F-4 Phantom IIs in formation, prior to delivery to their respective air forces. From the top, an RF-4B of the USAF, an F-4M of the RAF, and a RF-4C of the Luftwaffe. The longer nose of the RF versions is a distinguishing feature. (Photo: McDonnell Douglas.)

20 Another of tomorrow's aircraft, the Dassault-Breguet-Dornier Alpha Jet advanced jet trainer, which will be in service with the Luftwaffe and the Armée de l'Air during the mid-1970s. It will replace the Potez Super Magisters of the Armée de l'Air, and the Lockheed T-33As and Cessna T-37s of the Luftwaffe. (Photo: Dornier.)

21 Operating from remarkably pleasant surroundings are these Dornier Do 27 liaison aircraft of the Heeresflieger (Federal German Army Aviation). These aircraft have a very good short-field performance, and can accommodate up to five persons including the pilot. (Photo: Ministry of Defence, Bonn.)

22 Very much aircraft of today – two Lockheed F-104G Starfighters of the Luftwaffe prior to a training flight. Although basically an interceptor, the European-built versions of the Starfighter have been developed into multi-mission aircraft which can undertake strike duties as well. (Photo: Ministry of Defence, Bonn.)

23 An Indian Air Force Hawker Siddeley Gnat fighter pilot operates his undercarriage, probably to help reduce his speed to that of the photographic aircraft! A number of Gnats were built under licence in India by Hindustan Aircraft, and proved themselves to be exceptionally manoeuvrable aircraft during conflict with Pakistan Air Force North American Sabres, whose air-to-air missiles could not function properly at low height under tropical conditions (due to intense heat rising from the ground). Nicknamed the 'Sabre-slayer', a further two hundred Gnats were put into production. The other user of the Gnat was the Finnish Air Force, while the RAF used two-seat versions of this British-designed aircraft for advanced training. (Photo: Indian Air Force.)

24 The Indian Air Force has three squadrons of BAC Canberra B(1)58 intruder bombers, which are a version of the standard B(1)8 used by the RAF for many years. One of the Indian aircraft is shown here, at some considerable height. (Photo: Indian Air Force.)

25 Nine Fiat G.91s of the Italian Air Force aerobatic team, 'Le Frecce Tricolou', taxi out before a display. The Fiat G.91 is the Italian Air Force's standard light fighter-bomber, and has also been operated by the Luftwaffe. (Photo: Aeronautica Militara Italiana.)

26 An Agusta-Bell 204B Iroquois helicopter of the Italian Air Force. This is just one of a number of Bell designs built under licence in Italy by Agusta. The Iroquois has also been built under licence in Japan and West Germany. (Photo: Aeronautica Militara Italiana.)

27 The NAMC XC-1A STOL tactical jet transport for the Japanese Air Self-Defence Force. This aircraft can accommodate sixty fully-equipped troops, and production deliveries of forty will start in 1973. Another NAMC aircraft in Japanese service is the YS-11 turboprop tactical transport. (Photo: Japanese Self-Defence Agency.)

28 A flight of Japanese-built Lockheed F-104J Starfighter interceptors of the Japanese Air Self-Defence Force. The JASDF received more than two hundred of this aircraft from the Japanese aircraft industry. Those shown are armed with Sperry Sparrow air-to-air missiles. (Photo: Lockheed.)

29 A solitary BAC 167 Strikemaster armed-jet trainer of the

Kuwait Air Force. This aircraft is a development of the Jet Provost and is used by a number of air forces for counter-insurgency (COIN) and light strike operations. An unusual feature for a warplane, shared with the BAC 145 jet trainer, is a pressurized cockpit. The aircraft shown is equipped with two 75-gallon long-range fuel tanks. (Photo: BAC.)

30 The Royal Malaysian Air Force operates sixteen of these Sikorsky S-61A helicopters in the transport, search and rescue roles. (Photo: Sikorsky.)

31 A Breguet Br. 1150 Atlantic maritime-reconnaissance aircraft of the Royal Netherlands Navy. These aircraft are operated from shore bases over the North Sea, and replaced Lockheed Neptunes. (Photo: Royal Netherlands Navy.)

32 Two of the twelve Fokker F-27M Troopship transports of the Royal Netherlands Air Force. There are a number of the military version of the highly successful Friendship airliner in service around the world. (Photo: Royal Netherlands Air Force.)

33 The standard ground-attack fighter of the Communist bloc countries is the Sukhoi Su-7. The Polish Air Force has some four wings of the Su-7B type, with one flight shown flying in formation here. (Photo: Polish Air Force.)

34 Two Mikoyan MiG-21PF 'Fishbed D' all-weather interceptors of the Polish Air Force. The under-fuselage air brakes of this type can be seen clearly on the lead aircraft. Unarmed, the two shown appear to have reconnaissance pods fitted. (Photo: Polish Air Force.)

35 Although largely replaced in Polish Air Force service by Su-7Bs, a few Mikoyan MiG-17 fighters remain, adapted for the ground-attack role, while some Communist bloc air forces still have quite large numbers of this type, which was basically a development out of the MiG-15 (Photo: Polish Air Force.)

36 The South African Air Force uses some eighty Atlas Aircraft-built Aermacchi MB.326K jet trainers for training purposes, and this aircraft is known as the Impala in SAAF service. Another two hundred or so armed versions are used by the Citizen

Defence Force, whose North American T-6 Harvards they replaced. (Photo: South African Air Force.)

37 In 1959 the South African Air Force took delivery of eight Avro (now Hawker Siddeley) Shackleton MR.3 maritime-reconnaissance aircraft, and these still serve in an anti-submarine squadron. Since this aspect of the SAAF's duties is now of vital importance, with the increased use of the Cape route by shipping following the closure of the Suez Canal in 1967, it is likely that a similar number of Hawker Siddeley Nimrod aircraft will be ordered soon. (Photo: South African Air Force.)

38 A small number of Lockheed C-130B Hercules transports operate alongside Transall C.160 and Douglas C-47 Dakota transports in the South African Air Force. (Photo: South African Air Force.)

39 Rarely photographed, although very important is the III-B two-seat conversion trainer version of the Dassault Mirage. Shown here is a Dassault Mirage III-BZ of the South African Air Force. (Photo: South African Air Force.)

40 The Swiss Air Force and Anti-Aircraft Command has two squadrons operating a total of thirty-six licence-built Dassault Mirage III-S interceptors, and one squadron with eighteen Swiss-built Mirage III-RS reconnaissance-fighters. One of the latter is shown here in typical operating conditions preparatory to take-off. (Photo: Swiss Air Force.)

41 A Russian Tupolev Tu-20 'Bear' turboprop bomber aircraft being escorted away from land by a General Dynamics F-102 Delta Dagger interceptor of the USAAF, operating from NATO base in Iceland. (Photo: Central Office of Information.)

42 A Russian Independent Navy Air Fleet Kamov Ka-25 'Hormone' shipborne helicopter flying past the British aircraft carrier HMS *Ark Royal*, from whose deck this photograph was taken. The Ka-25 is Russia's standard naval helicopter. (Photo: Central Office of Information.)

43 Middle Eastern line-up. BAC Lightning F.53s, export versions of the F.3 in RAF service! for the Kuwait Air Force, those nearest to the camera; and for the Saudi Arabian Air Force, the three furthest away. These differ from the RAF versions inasmuch as the RAF's Lightnings are interceptors, while the export versions are multi-mission aircraft capable of strike duties. Both Kuwait and Saudi Arabia are receiving T.55 conversion trainers, in turn export versions of the RAF's T.5s, in addition to the combat types shown. (Photo: BAC.)

44 The mid-1970s will see the RAF and Luftwaffe each receiving some four hundred and the Aeronautica Militara Italiana one hundred of the Panavia 200 Panther multi-role combat aircraft; an artist's impression of this twin-engined variable-geometry type is shown here. Manufacturers are BAC, Messerschmitt-Bölkow-Blohm and Fiat. (Photo: BAC.)

45 The RAF is receiving 164 BAC-Breguet Jaguar tactical strike aircraft during the early 1970s, plus thirty-six two-seat conversion trainers. One of the strike versions is shown here armed with bombs and rockets, and two cannon. (Photo: BAC.)

46 A Hawker Siddeley Harrier conversion trainer of the RAF shown while taking off vertically. Noticeable features are the enlarged cockpit for two seats, extended tail, and the camouflage markings now used by the RAF's Harriers, on which the white is omitted from the red, white and blue roundels and tail flashes. (Photo: Hawker Siddeley.)

47 The Hawker Siddeley Hunter was one of the most successful post-war British aircraft, and was built under licence in a number of countries. Demand for refurbished second-hand aircraft runs at a higher level than supply. Only a few remain in the RAF after replacement by Harriers and Jaguars. Shown here is an RAF example. (Photo: Hawker Siddeley.)

48 The RAF uses thirty-eight Hawker Siddeley Nimrods, maritime-reconnaissance developments of the Comet airliner, with the Comet's Avon turbojets replaced by Spey turbofans and other modifications including a double-bubble fuselage. The South African Air Force is a prospective user of this aircraft, which is the world's first jet maritime-reconnaissance type. (Photo: Hawker Siddeley.)

49 A Westland SH-3D Sea King helicopter dunks its sonar equipment. The Westland Sea King is a developed version of the Sikorsky S-61 SH-3D Sea King, and is fitted with British engines and other equipment. Its anti-submarine detection apparatus is as extensive as that of a Leander class frigate (although the helicopter's four-hour endurance is far less than that of the ship), with the advantage of no propellor noise under water to distort reception. The Fleet Air Arm has sixty Sea Kings. (Photo: Westland.)

50 A Westland Wasp helicopter takes off from the landing pad of a Royal Navy frigate. These small helicopters are deployed aboard British frigates for anti-submarine duties. The vessel's own depth-charge launchers can be seen below and behind the helicopter. (Photo: Westland.)

51 The British Army uses a large number of Westland-Bell 47 Sioux helicopters for liaison and AOP duties. The flight shown here, however, are an aerobatic team, the Blue Angels. (Photo: Army Aviation.)

52 A British Army Westland Scout helicopter fires an SS 11 wire-controlled anti-tank missile. This helicopter is used for attack and liaison duties mainly, with some light transport and communications work thrown in. The similarity in appearance to the Wasp is not surprising, and is intentional. Both helicopters use in fact more than sixty per cent common components. (Photo: British Army Aviation.)

53 The Westland design office's share of the Anglo-French helicopter package deal is the WG.13. This will be for the British Army (as with the cardboard and wood mock-up shown here), the Royal Navy and RAF, with plans for French Army and Navy use too. A number of other navies are also interested. The Westland Scout and Wasp will be replaced by the WG.13. (Photo: Westland.)

54 The McDonnell Douglas F-15 is due to be in USAF service by the mid-1970s. It will be an air-superiority fighter to counter Russia's MiG-23 'Foxbat' and, as will be seen from this artist's impression, it will have a double fin, a feature also of the MiG-23. (Photo: McDonnell Douglas.)

55 The world's first operational variable-geometry aircraft was the General Dynamics F-111A tactical fighter of the USAF. shown here. Bomber versions, FB-111, are now in service with the USAF. An Australian order is still very much in the balance, however, while a British order for the F-111K has been cancelled. (Photo: USAF.)

56 The world's largest transport aircraft is the Lockheed C-5A Galaxy, with twice the carrying capacity of a 'Jumbo' jet. The USAF has some eighty-five of these aircraft, and one is shown here. (Photo: Lockheed.)

57 A Northrop F-5A Freedom Fighter, nearest the camera, and a Northrop T-38 Talon advanced trainer are shown here taking off side by side. So far, the USAF and Luftwaffe are the only operators of the T-38, but F-5As and the two-seat F-5B are in service with many air forces in Europe, Africa and Aisa. (Photo: Northrop.)

58 A Cessna A-37 on an attack mission in South Vietnam fires its rockets. The A-37 is the attack version, or armed-trainer, of the T-37 trainer. It is used in South Vietnam by the USAF and the South Vietnamese Air Force. The aircraft illustrated belongs to the USAF. (Photo: Cessna.)

59 In the air over Vietnam. A flight of Douglas A-1E Skyraiders is followed by a Sikorsky CH-3B helicopter which is being refuelled in flight by a Lockheed KC-130H Hercules tanker. The flight refuelling equipment is positioned under the wings, and the booms are black with white rings for easy identification. KC-130Hs have to be used for helicopter refuelling instead of the KC-135, which is the standard USAF tanker, because of the lower speeds involved ruling out the use of jet tanker aircraft. The helicopter is available for rescue duties should one of the Skyraiders be forced down, and the availability of a tanker vastly extends the helicopter's radius of action. (Photo: Sikorsky.)

60 A scene back at base in South Vietnam. A Cessna O-2A of the USAF is in the middle of the picture, with a Bell UH-1H Iroquois and a Sikorsky S-58 helicopter in the left and right

background respectively. The O-2A is a military version of the Cessna Super Skymaster executive aircraft, and is used for forward air control of ground-attack missions, liaison and AOP, tactical reconnaissance and light attack duties, on which latter mission the aircraft in the photograph is being prepared, armed with rocket pods. A feature of the O-2 and its civil cousin is the use of two engines, one with a conventional propeller, the other, between the twin-booms, using a pusher-propeller. (Photo: Cessna.)

61 Still in Vietnam. A USAF Sikorsky CH-3C takes a light artillery gun to the battlefield. This helicopter, another version of the S-61, can also carry up to thirty passengers. (Photo: Sikorsky.)

62 A Ling-Temco-Vought A-7E Corsair II attack aircraft of the United States Navy above the attack carrier USS *Ranger* off the South Vietnamese coast. The LTV A-7E Corsair II is a Rolls-Royce Spey-powered version, with the engine built under licence in the USA by Allison, and it is one of the most effective aircraft ever to enter USN service. Earlier versions of the Corsair II, the A-7A-Cs, used American engines. (Photo: United States Navy.)

63 Also off the South Vietnamese coast was the USS *Constellation* at the time this photograph was taken of a USN McDonnell Douglas F-4B Phantom II. The aircraft has just taken off using the angled flight deck and waist catapult. (Photo: United States Navy.)

64 Another shot from the USS *Constellation*. This time a Grumman E-2 Hawkeye airborne-early-warning aircraft comes into land, guided by the landing-signal officer who is using a deck console of instruments and a radio telephone to guide the aircraft in. The arrester hook can be seen clearly in the 'down' position. The Hawkeye replaced the Grumman E-1 Tracer aboard the attack carriers, although the anti-submarine carriers, which are smaller, still use Tracers. (Photo: United States Navy.)

65 A United States Navy maritime-reconnaissance Lockheed P-3 Orion while on patrol flies low over a United States Coastguard cutter. The turboprop Orion replaced the famous Lockheed P-2 Neptune. Amongst operators of the Orion are the United States Navy, RAAF, and RNZAF, and the Canadian Armed Forces, who are the latest customer and who will have thirty built under licence in Canada to replace their Canadair Argus aircraft. The US Coastguard uses Lockheed Hercules aircraft for maritime-reconnaissance and rescue duties. (Photo: United States Navy.

66 The standard search and rescue and plane-guard helicopter for the United States Navy is the Kaman UH-2 Seasprite, shown here. Plane-guard duties consist of keeping a helicopter or warship positioned astern of an aircraft carrier while flying is in progress, in order to rescue the crew of any aircraft which crashes into the sea. (Photo: Kaman.)

67 The United States Marine Corps uses these Bell AH-1J Sea Cobra helicopters for attack duties. Basically this is a twin-engined version of the Bell AH-1 Hueycobra, in turn a gunship development of the highly successful Bell UH-1 Iroquois. (Photo: Bell).

68 The Hawker Siddeley Harrier vertical take-off strike aircraft is currently entering service with the United States Marine Corps as the AV-8. Initial aircraft have been produced by the manufacturer, but, although further supplies from the same source have been provided, the possibility still exists of licence-production by McDonnell Douglas. The aircraft shown is carrying bombs, while the 'bulges' on either side of the bomb under the fuselage are cannon. The USN may also order Harriers in due course. The USMC aircraft have more powerful versions of the engines used in the RAF's Harriers. (Photo: Hawker Siddeley.)

69 The McDonnell Douglas F-4 Phantom II supersonic attack aircraft is appearing in increasing numbers. Phantom IIs are in service with the USAF, USN, USMC, the RAF and Royal Navy, RAAF, Luftwaffe, Republic of Korea Air Force, Imperial Iranian Air Force, and Israeli Air Force. The aircraft shown is one of the RF-4Bs which are replacing the three squadrons of LTV RF-8A Crusaders employed by the United States Marine Corps for reconnaissance duties. (Photo: McDonnell Douglas.)

70 The United States Army's standard medium-lift helicopter is the Boeing-Vertol CH-47 Chinook, which can carry up to forty-five fully-equipped troops or eight tons slung underneath the fuselage. Chinooks have been used extensively in South Vietnam. (Photo: United States Army.)

Major Defence Agreements and Alliances

Warsaw Pact

The Warsaw Pact is a multilateral military alliance formed in 1955, and at present the members are the Soviet Union, Albania, Bulgaria, Czechoslovakia, East Germany, Hungry, Poland and Rumania. China has never been a Warsaw Pact member, and nor has Yugoslavia. Albania was a founder member, but withdrew from membership in 1968 after eight years of non-participation. A number of aid and other treaties and pacts exist within the Warsaw Pact, which is controlled by its Joint High Command from Moscow. Albania is the only country to have been able to leave the Pact, and doubtless this was due to China's backing for the move, and the difficulty of mounting a military operation in some of the most difficult terrain in Europe. There has been no hesitation in using Warsaw Pact forces to enforce strict Communist control of Hungary and Czechoslovakia, although in the latter case there had been no intention of withdrawal.

North Atlantic Treaty Organization

The North Atlantic Treaty Organization, or NATO, dates from 1949, and is a multilateral military alliance which has a membership including Belgium, Canada, Denmark, France, West Germany, Greece, Iceland, Italy, Luxembourg, the Netherlands, Norway, Portugal, Turkey, the United Kingdom and the United States. Greece and Turkey joined in 1952, followed by West Germany in 1955, while France is no longer a fully active member and Canada has been taking a lukewarm attitude to NATO for some years now. Control is through the North Atlantic Council which is based in Brussels (although originally the Council was based in Paris).

South-East Asia Treaty Organization

The South-East Asia Treaty Organization, or SEATO, dates from 1955, and is a multilateral military alliance with a membership including Australia, France, New Zealand, Pakistan, the Philippines, the United Kingdom and the United States, but without any formalized command structure. Apart from SEATO, there are other treaties, notably ANZUS, binding the United States, Australia and New Zealand together, and ANZAM binding Australia, New Zealand and United Kingdom forces in Malaysia, while a separate Anglo-Malaysian treaty also exists.

Central Treaty Organization

The Central Treaty Organization, or CENTO, is a multilateral defence treaty with a membership including Iran, Pakistan, Turkey and the United Kingdom, with the United States as an associate member. In common with SEATO, there is no formalized command structure. Originally CENTO was known as the Baghdad Pact, but a revolution in Iraq led to the severing of ties between that country and the other members.

NATO Identification Names given to Russian Aircraft

Since its inception, the North Atlantic Treaty Organization has followed the practice of allocating identification names to Soviet aircraft, starting with 'F'–fighter, 'B'–bomber, 'C'–transport 'H'–helicopter, 'M'–maritime, trainer, or reconnaissance, and with A, B, C, etc. suffixes for developments of a basic type. This system has the advantages that any new Russian type can be readily described by its identification name, and is superior to a system of identification numbers in that no confusion relating to the order of introduction or appearance results. The design bureau name and design number, the Russian equivalent of manufacturers' names and type numbers, is often not known for some time after the first sighting of a Soviet aircraft type.

The identification names are given below:

Fighters
Mikoyan MiG-15 'Fagot'   Mikoyan MiG-17 'Fresco'
Mikoyan MiG-19 'Farmer'   Mikoyan MiG-21 'Fishbed'
Mikoyan MiG-23 'Foxbat'   Mikoyan MiG- 'Flogger'
Sukhoi Su-7 'Fitter'   Sukhoi Su-9 'Fishpot'
Sukhoi Su-11 'Flagon'   Tupolev Tu- 'Fiddler'
Yakovlev Yak-25 'Flashlight' Yakovlev Yak-28 'Firebar'
Yakovlev Yak- 'Freehand'

Bombers
Ilyushin Il-28 'Beagle'   Myasishchev Mya-4 'Bison'
Tupolev Tu-14 'Bosun'   Tupolev Tu-16 'Badger'
Tupolev Tu-20 'Bear'   Tupolev Tu-22 'Blinder'

Maritime
Beriev Be-6 'Madge'   Beriev Be-10 'Mallow'
Beriev Be-12 'Mail'

Reconnaissance
Yakovlev Yak- 'Mandrake'

Transports
Antonov An-2 'Colt'
Antonov An-14 'Cold'
Antonov An-24 'Coke'
Ilyushin Il-18 'Coot'
Tupolev Tu-104 'Camel'

Antonov An-12 'Cub'
Antonov An-22 'Cock'
Ilyushin Il-14 'Crate'
Lisunov Li-2 'Cab'
Tupolev Tu-114 'Cleat'

Helicopters
Mil Mi-4 'Hound'
Mil Mi-8 'Hip'
Kamov Ka-25 'Hormone'

Mil Mi-6 'Hook'
Mil Mi-10 'Harke'

Trainers
Ilyushin Il-28U 'Mascot'
Mikoyan MiG-21UTI 'Mongol'
Yakovlev Yak-11 'Moose'

Yakovlev Yak-25 'Mangrove'
Mikoyan MiG-15UTI 'Midget'
Sukhoi Su-7UTI 'Moujik'
Yakovlev Yak-18 'Max'

Abbreviations used in the Text

In most cases abbreviations appear only in individual entries where their meaning is immediately apparent—for example, a reference to RAF in the Royal Air Force entry—and need no explanation. Those following are used in less obvious contexts.

Aèronavale–French Navy Air Arm
AOP–air observation post
L'Armée de l'Air–French Air Force
ASW–anti-submarine warfare
BAC–British Aircraft Corporation
CAF–Canadian Armed Forces
COD–carrier onboard-delivery
COIN–counter insurgency
FAA–British Royal Navy's Fleet Air Arm
IAF–Indian Air Force
ICBM–inter-continental ballistic missile
JASDF–Japanese Air Self-Defence Force
JGSDF–Japanese Ground Self-Defence Force
JMSDF–Japanese Maritime Self-Defence Force
Luftwaffe–Federal German Air Force
NATO–North Atlantic Treaty Organization
RAAF–Royal Australian Air Force
RAF–British Royal Air Force
RCAF–Royal Ceylonese Air Force
RDAF–Royal Danish Air Force
RHAF–Royal Hellenic Air Force
RMAF–Royal Malaysian Air Force
RNAF–Royal Netherlands Air Force
RNZAF–Royal New Zealand Air Force
RSwAF–Royal Swedish Air Force

STOL–short take-off and landing
USAF–United States Air Force
USAAF–United States Army Air Force
US Army–United States Army
USMC–United States Marine Corps
USN–United States Navy